The Adriatic Islands and Corfu

Celia Irving

The Adriatic Islands and Corfu

J M Dent & Sons Limited London

First published 1971

Made in Great Britain at the
Aldine Press Letchworth Hertfordshire
for J M Dent & Sons Limited
Aldine House Bedford Street London

ISBN 0 460 03922 9

For the men in my life; husband, sons, — and not forgetting my father — I have written this book about the blue Adriatic and its islands where the women are gentle and still remember that theirs is a Man's World.

Contents

Illustrations

Between pages 84 and 85

Introduction

All the Adriatic islands, except Corfu to the east, lie off the Yugoslav coastline of Istria, Dalmatia and Montenegro. Floating as they do in the cradle of the Western world's classical civilization, the islands are historical treasure-houses of artistic, archaeological and architectural interest. As islands they are quite young; geographically speaking about 30,000 years, though their geological content is much older. Originally they were the tops of mountains similar to those near the coast on the mainland. But the coastal area became submerged under the sea at the end of the last Ice Age, during the biblical flood, leaving only the peaks above water and turning escarpments, chimneys and other formations usual in limestone mountains into the beautiful bays and rocky inlets, wooded slopes and white cliffs which make up their present fabulous shores. There are, in all, some 1,040 islands if you include the small rock-reefs. They all bask in the blue Adriatic, are assured of brilliant hot sunshine in summer and a temperature which hardly ever falls to zero in winter. The scenery and colours, with the dark green of pine trees, and the sea which shades from blue to emerald near the shore, contrast arrestingly with the white or rose-pink of the rocks and beaches and the honey-coloured stone of the small medieval walled towns. Paradise enough for the lotus-eaters; but the islands have more to offer than just beauty for anyone who wishes to dig deeper.

Every year increasing numbers of people visit these beautiful islands, and the habit of making one of them the focal point from which to visit the mainland and other islands is becoming more and more popular. Of course, they are all enchanting with their felicitous climate and their vivid scenic beauty, but what is not generally known is their extraordinary individuality, both of terrain and of people. Each of them has a closely knit community differing from the next, though they may be only a few kilometres apart. History and geographical location have formulated a different character for each. For example, the people of Hvar and Korčula, only thirty sea miles apart, or Lošinj and Susak, nine miles apart, are as different in character and manner as the Scots and the Cornish, or the Dutch and the French. Even their dialects vary so much that it is difficult for some Yugoslavs to understand them completely.

Because of their special location in the coastal shipping lanes which join the western with the eastern world, the Adriatic islands have always enjoyed a front-row seat in history, and even prehistory, which obligingly unfolded itself on their very shores. Quite often the islanders have left their front stalls and joined in the drama themselves. Sometimes the action concerned the battles and schemes of the ancient Greeks and Romans; sometimes the patchwork intrigues of the Venetians and the Genoese; the mainland pressures of Ostrogoths, Franks, Hungarians and Austrians; or the strategies and raids of the British against their common enemy the Napoleonic French. Later, in company with the partisan forces during the struggles of the Second World War, British and Slavs joined forces yet again. Greeks, Romans, Byzantines, French, Austrians and Italians — all have left their mark on the history of the Adriatic islands, yet no foreign yoke has managed much to change the basic character of the people. They all retain the essentially independent traits of their ethnic group.

The original inhabitants of the Adriatic were Stone Age cave-dwellers. The Illyrian tribes emerged during the Metal Ages — Copper, Bronze and Iron. Copper appeared about 4,000 years ago. These Illyrians have left their tribal names in some areas, such as Histris in Istria, Dalmatians in Dalmatia, and Liburnians in Liburnia, the coastal belt near Rijeka. The ancient Greeks arrived in the Adriatic to colonize the rich trade and Amber route. This route had existed in prehistoric times, and joined east and west with the Baltic, bringing with it the prized amber and its supposed supernatural qualities. They drove the Illyrians away from some of the best shore-side centres, but they did not eliminate the Illyrian strain. The Illyrians moved to the hills, and there they lived on. The same thing happened during the Roman invasion, and it was only later, towards the end of the Byzantine Age, that these often blue-eyed and red-haired people were forced to move eastwards, many to settle eventually in the area which is now Albania.

So, though Greeks and Romans settled like a veneer, the Adriatic people were basically Illyrian and provided many outstanding leaders, such as the Emperor Diocletian, perhaps the greatest Dalmatian-Illyrian of them all. It was not until the great migration of the Slav peoples in the sixth and seventh centuries that basic ethnic qualities began to change. The Slavonic ethnic group of Indo-Europeans probably originated between the Carpathian mountains and the Dnieper river in prehistoric times, but very little is known for certain. When the peaceful Slavs, a hundred years after the rampaging Avars, came to Dalmatia, they brought their families and their tribal farming customs and settled in quietly beside what remained of the artistic Illyrians. The two peoples probably had many similar characteristics. According to a document left by Maurice, Roman Emperor of the East, and quoted in Zvane Črnja's *Kulturna Historija Hrvatske*, the Slav tribes were free and could not be induced to live in slavery or subjection; they were particularly brave and persistent when fighting in their homeland. To the stranger they were kind, accompanying him willingly to his place of destination and protecting him. The women, devoted to their husbands, often gladly killed themselves rather than face widowhood. Another writer, Procopius, stated that they were not ruled by one man alone but, since ancient times, had lived under popular government; and fortune and misfortune were shared alike by the whole community. Early witnesses to Slav hospitality and steadfastness which still pertain today! Peace-loving and slow to anger, they still undoubtedly are. But, once provoked beyond a certain point, they become dogged and determined fighters, capable of great sustained effort.

For 1,500 years civilization has continued to evolve in southern Europe, uninterrupted by any 'dark age' after the fall of the Roman Empire, and sometimes in spite of foreign domination. And there are instances of a very early social conscience in the region. Perhaps it is the result of many examples of enlightened thinking, together with an objective view and acceptance of even the harshest treatment, which makes the Adriatic people seem so adult in their outlook on life; an objectiveness which allows them a simplicity of feeling for the fundamental things that count. One can discern it in university graduates and illiterate villagers alike; and, whatever it is, the charm for the visitor it happens to touch is unforgettable.

It is often said that the Slavs like a good intrigue, much as they enjoy a good game of chess. This is, perhaps, an over-simplification. But they have a certain tough, adult way of finding a sardonic amusement in every situation. Though normally kind-hearted, their burning steadfastness has, from time to time, led them to acts of great cruelty against those considered to have broken their ethical laws. In the past, women were stoned and outcast for infidelity; men of their own country maimed and tortured for supporting the wrong cause. At the same time men, both in the middle ages and in the twentieth century, have suffered these cruelties with comprehension, whether they were seventeenth-century noblemen, would-be reformers, or the guerrilla fighters of the last war.

Through the ages the islands have received refugees from the mainland escaping from Franks, Saracens, Mongols, Turks, and lastly Germans. Conversely, each war depleted island numbers; in the Second World War alone they lost 10,000 people. Now the younger men and women tend to seek work on the mainland. But, speaking to young people in Zagreb in October 1969, President Tito urged that graduates should know they must not only work in the big towns but consider also the small places which need their skill, and which have just as much to offer them socially, politically and culturally.

For me, the islands are at once so different in atmosphere, so small in their intimate and insular relationships, so enveloping in their eager, vivid affections and so intense in the details of their everyday life, that personal encounters become inextricably mixed with overall impressions, and particular anecdotes are bound to intermingle with the broad sweep of factual narrative. For such has been the human evolution of these classic lands that personal contact and individual reaction to people and events are the motivations which have brought them, as they are, to the twentieth century.

Acknowledgments

I should like to thank all those who have so generously given me help and information while I have been writing this book. In particular: Rudi Pisk, Director of Yugotours; Nikša Kuiš, Commercial Director, Jadrolinija Shipping Line, Rijeka; John Damaschinos, until July 1969 Director of Tourism, Corfu, Greece; Zlatko Uzelac, Dejan Živojinović and Mile Janković of the Yugoslav National Tourist Office, London; Basil Iatridis, Director National Tourist Organization of Greece, London; Milan Zupan, Yugoslav Embassy, London. The tourist organizations of Zagreb, Rijeka, Istria, Zadar, Šibenik, Split, Dubrovnik, Montenegro and Corfu; together with all those who have afforded unfailing help, hospitality, welcome and friendship on all my visits to the islands of the Adriatic.

Note on pronunciation

Nothing is more irritating for a reader than being stopped short in the middle of a narrative by what looks like an unpronounceable name; similarly, for anyone conversant with Slav languages, there is nothing so annoying as an anglicized spelling of the same name. It is a cleft stick. Fortunately, though it sometimes appears almost impossible at first sight, the Yugoslav language is not difficult to pronounce if a few rules are borne in mind, though purists might call the following an over-simplification!

Every letter is pronounced; vowels never vary and are monosyllabic pure sounds (u is invariably oo); dj (or Đ, đ) is like g for George; j is like y; c like ts; and, whenever there is a circumflex accent, you insert an 'h'; while an acute accent on a 'ć' turns it into 'tiu' like the 't' in future.

In Montenegro, Serbia and Macedonia, the Cyrillic alphabet is normally used, but the same simple rules of pronunciation still obtain.

Note: Sv. is short for Sviti or Sveta, meaning Saint.

Date lines

B.C.		
150,000	Neanderthal Man.	Low forehead, thick brow, sloping jaw. Stone scrapers, knives. Could make fire. Tended sick. Disappeared into the Ice Age.
50,000	Old Stone Age and Paleolithic.	Traces of lions, mammals, etc. in Europe.
10,000	Mesolithic Man.	From Drava valley to Adriatic.
3,000	Neolithic Man.	Women could make pottery.
2,000	Copper Age.	Eneolithic, when stone and copper were hand-in-hand. Amber route via Danube from Baltic to Adriatic.
1,500	Bronze Age.	Illyrians learnt to mix metals — Indo-European Illyrians.
1,000	Iron Age.	Greek zenith of civilization.
734	Corfu (Corcyra) founded by Corinthians.	
400	New Iron Age.	(La Tène.)
c. 397	Vis colonized.	With help of Dionysius of Syracuse.
384	First battle for Hvar.	
3rd C.	Apollonius of Rhodes.	Writing end of 3rd and beginning of 2nd centuries B.C.

B.C. during 3rd C.	Queen Teuta fights back	after death of her husband Agron.
229	Romans took Corfu (until A.D. 733).	
228 and 219	Roman expeditions in Adriatic.	
177	Brioni islands came under Roman rule.	Nesactium in Istria also fell.
27 B.C.–14 A.D.	Reign of Caesar Augustus.	
11	Caesar Augustus moved Italian border.	To near the east coast of Istria. Istria became part of the Xth Italian Region, Venetia et Histria.
A.D. 9–11	Tiberius annexes Adriatic.	16.1.12 celebrated 'Triumph' in Rome.
254–313	Diocletian.	
5th C.	Visigoths, Huns and Ostrogoths.	
535	Justinian reconquered coast.	
582–602	Maurice, Byzantine Emperor.	
Mid 6th C.	Avars appeared from Danube.	Indo-European ethnic group. 550–1 spent winter on Adriatic coast (Procopius), then disappeared.
6th and 7th centuries	South Slavs arrived.	Peacefully invaded and settled.
768–814	Charlemagne King of Franks.	Holy Roman Emperor 800. Empire spread almost to Split, coast left to Byzantium. Majority of Illyrians moved eastwards, eventually settling in what is now Albania.
827–869	St Cyril (and fellow Brother Methodius).	Invented Cyrillic alphabet about 860–70.

10th C.	Magyars (Finnish-Ugrian tribes) came to the Danube valley from east and north of Bulgaria.	This migration separated the South Slavs from Czechoslovakia, Poland and Russia; South Slavs developed on separate lines.
928	Bishop Grgur of Nin lost bishopric.	
924	Tomislav assumed title of king.	Crowned at Nin; he was a Župan, or chieftain, and the first to assume the title of king.
911–959	Constantine VII (Porphyrogenitus)	born 'in the purple' (when father was on the throne — in a chamber called the 'Porphyra' (purple). Man of letters.
1000	Orseolo of Venice appeared.	Doge Peter Orseolo made forays into Dalmatian area.
1058–1074	Petar Krešimir.	Strong ruler and Croat hero. 'Golden Age'.
1076–1089	Dmitar Zvonimir, King of Croatia.	Crowned by legate of Pope Gregory VII, Monseigneur Gebison, and promised him 200 gold bezants annually. Married Helen, daughter of Ladislas I of Hungary. Zvonimir, a papist, was looked on with disgust by the Slav heretic faction whose claimant was Slavac. He mysteriously 'disappeared' of course!
1077–1095	Ladislas I of Hungary (1091–1095 Ladislas ruled in Croatia).	
1102–1108	King Koloman, king of Hungary took over Croatia.	Koloman was king of Hungary, 1095–1114. Crowned king of Croatia 1102. Beginning of long association between Croatia and Hungary. Koloman gave Croatia autonomy, domestically, under a Ban; sometimes 'noble', sometimes royal.

1114–1131	Stepan II followed Koloman.	From this time Croatia and Hungary marched together, although Dalmatia (and parts of Dalmatia) changed hands frequently — always bickering with Venice.
1123	Saracens overran Trogir.	
1176	Goldsmiths' Guild already in existence in Zadar.	
1177	Pope Alexander III went on 'progress'.	Visited Rab and Zadar.
1196	Serbian king Stepan Nemanja reached the Adriatic coast.	From Neretva river to Kotor and Bar.
1202	Fourth Crusade diverted to subdue Zadar.	
1203–1235	Andrew II of Hungary.	He called himself hereditary king of Hungary, Dalmatia, Croatia, Serbia, etc.
1235–1270	Bela IV, son of Andrew II.	Bela IV was harassed by the Mongolian hordes of the ruler Ogodai, son of Genghis Khan. See Split and Trogir.
1266–1285	Charles (of the house of Anjou)	reigned in Naples and Sicily.
1290–1295	Charles Martel (of the house of Anjou).	Invaded Croatia and had himself crowned at Zagreb.
1282–1321 1331–1355 1355–1371	King Uroš II (Nemanjid). King Uros IV (Dušan). King Uroš V.	The famous Nemanja dynasty built up a solid state in Serbia between 12th C. and the death of Stepan Dušan (1331–1355).
12th–16th centuries	The Frankopans, princes of Krk.	Were powerful in Croatia and Dalmatia, like their friends the Zrinski family who came from the Šubić family.
1322–1353	King Stjepan Kotromanić of Bosnia.	He was a Bogomil (breakaway religion) but became R.C. in 1340. In 1325 acquired the land of the Zahumlje — later called Hercegovina.

1353–1391	Stjepan Tvrtko (tvrtko means tough or hard).	Strong Bosnian king. In 1376 became king of Serbs, Bosnia, and the coast (incl. some islands). 1390, became king of Dalmatia and Croatia.
1358	Treaty of Zadar.	In 1345 Zadar had put itself under Ludovik's protection. In 1358 — in the town of Zadar — Ludovik dictated terms to Venice in which the Doge gave up claims to Dalmatia and Zadar.
1376	King Ludovik I claimed the throne of Hungary, Croatia, etc., married Elizabeth, daughter of Bosnian Stjepan Kotromanić.	Ludovik (or Louis) was of the House of Anjou (1342–1382). A strong king.
1382	After death of Ludovik I Zadar showed allegiance to the German king Sigismund, but changed to Ladislas of Naples.	Sigismund (of Luxembourg) was known as king of the Romans.
1386	Venice took over Corfu (on request).	
1389	The Battle of Kosovo Polje.	The terrible Serbian/Turkish battle of great bloodshed in which the people of Trogir joined.
1403–1414	Ladislas was crowned in Zadar 'king of Hungary and Croatia'.	Ladislas was born in 1376 and was the son of Charles of Durazzo (murdered in Hungary 7.2.1386) who was of the male line of the House of Anjou. In some areas Ladislas was regarded as a new 'white hope' in troubled European affairs, but in Dalmatia he was looked on with dismay.
1409	Ladislas of Naples sold his Dalmatian rights to Venice.	Venice paid Ladislas 100,000 ducats.
1420	Venice arrived in strength throughout Adriatic.	

1459/63/83	Serbia, Bosnia, Hercegovina, fell to the Turks.	Venice and Turkey cheek by jowl in the Adriatic area.
15th–17th centuries	Uskok pirate/heroes	Worked Adriatic seaboard.
1526	Turks defeat Hungary.	Venice begins to feel pressure.
1540	The peace leaves Venice only seaboard.	Turkish Beg ruled interior from Klis, near Split.
1571	Battle of Lepanto.	Turkish naval power crushed.
1645–69	Intermittent war between Venice and Turkey.	Peace of 1669, Dalmatia stayed with Venice but Crete went to Turkey (ref. Visovac).
1671	The last of the Frankopans and Zrinski-Šubić were executed.	Prince Krsto Frankopan and Petar Zrinski (whose family came from the powerful Šubić tribe, Bans of Trogir, etc., in 13th and 14th C.s) were beheaded in Vienna.
1699	Treaty of Carlowitz.	Demarcation line gave all coast north of Pelješac (incl. Korčula) to Venice. Pelješac and south (incl. Lastovo and Mljet) stayed with Dubrovnik. Except for two corridors, Turkey lost all access to the Adriatic. Corridors remained until 1914–18 war.
1797	Advent of Napoleonic French in Dalmatia, and also Corfu.	
1805	Russia in Kotor and Hercegnovi.	
1807	Treaty of Tilsit.	
1808	Dubrovnik officially ceased to exist.	
1811	British naval force on Vis.	Later also on Hvar, Korčula and Lastovo.
1814	French withdrawn from Dalmatia.	

1814–15	Congress of Vienna.	Treaty ceded Dalmatia, with Dubrovnik, to Austria.
1914	Archduke Ferdinand of Austria assassinated.	At Sarajevo, Hercegovina. World War I.
1915	Treaty of London.	Lords Grey and Oxford, authors.
1920	Treaty of Rapallo.	Italy gained Istria, watershed of Julian Alps, Zadar, and islands of Cres, Lošinj and Lastovo.
6.4.41.	Germany invaded Yugoslavia.	
28.6.48.	Expulsion of Yugoslavia from Cominform.	Complete break with Russia.
1954	Settlement of Trieste Zones 'A' and 'B'.	

JULIAN ALPS
Trieste
ISTRIA
Rijeka
Krk
Senj
Brioni Islands
Pula
Cres
VELEBIT MTNS
YUGOSLAVIA
Pag
Lošinj
Nin
Zadar
Krka R.
DALMATIA
Dugi Otok
Čikola R.
Kornats
Šibenik
Trogir
Split
DINARIC CHAIN OF MOUNTAINS
Neretva R.
Šolta
Brač
Hvar
Svetac
Šćedro
Pelješac Peninsula
Jabuka
Korčula
Brusnik
Biševo
Vis
Sušac
Lastovo
Mljet
Dubrovnik
Kotor
Boka Kotorska
Pescara
Palagruža
ADRIATIC SEA
ITALY
Bari
Brindisi
N
0 50 100 Miles
0 50 100 150 Km
ISTRIA
Brioni Islands
Pula

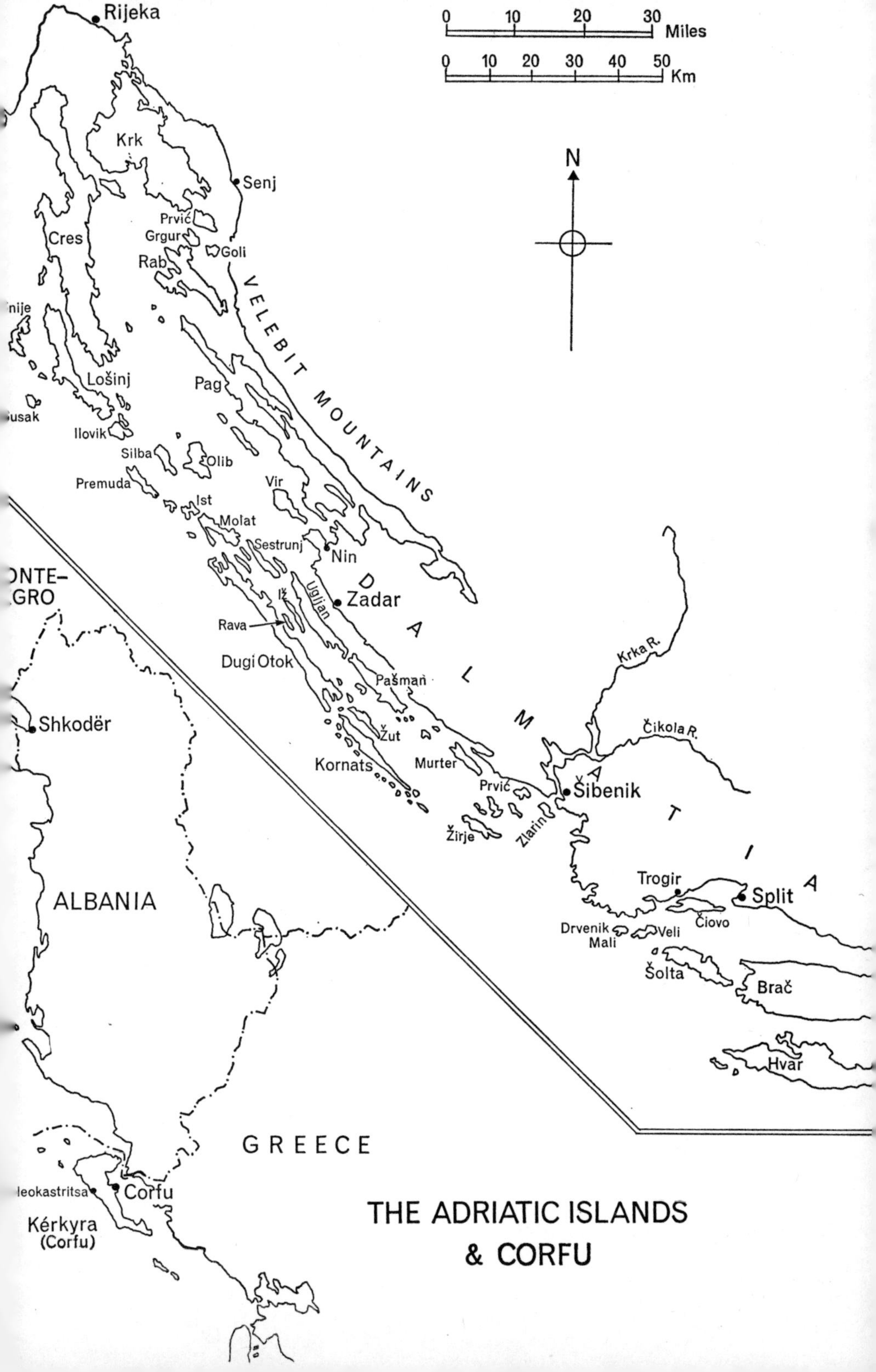

THE ADRIATIC ISLANDS & CORFU

Chapter 1 The islands of Brioni and Krk. The mainland towns of Pula and Rijeka

Travelling east along the Adriatic coast, the first islands of the long chain which extends to Greek Corfu are the Brioni group. The two major islands and a number of very small islets are situated outside the deep inlet of Pula and its large harbour. It is here that President Tito has a home to which he retreats with his wife whenever work permits. The Brioni islands have been inhabited since prehistoric times, were a favourite place for the well-to-do Romans, and have continued, by and large, to provide a peaceful retreat for various people through the ages. In its way, the largest island resembles an English estate or park, with rolling grassland interspersed with groups of tall trees. There are deer grazing in large herds and an abundance of hare, as well as pheasant, and other game birds, which lack the shyness of their British counterparts. The island is very small, only 690 hectares, and is crossed by a few asphalted drives; and ample Roman remains on Brioni point to the richness of Roman life in former times. In Verige bay are substantial traces of a luxury country villa of the first century. The grounds stretched for a kilometre by the sea and included a landing-stage and a breakwater. There were three temples, to Venus, Neptune and a third unknown god. Houses, farm buildings, thermae of considerable proportions, and the palace itself, built in tiers on the hill above the sheltered bay — the remains are all there. Near the top are two huge Roman water-cisterns, originally built to hold 25,000 cubic metres each.

A large defensive keep on Brioni was probably in use until the sixteenth century. Near by are the ruins of a sixth-century basilica dedicated to Sv. Marija, with the foundations of a Benedictine monastery deserted after an epidemic of plague in 1312; the basilica was reconstructed in the sixteenth century. Inside the church there is a round stone *Zadruga*[1] table — reminiscent of our King Arthur's — and a copy of the famous mural, *The Dance of Death*, by the Istrian painter Vincent of Kastva, painted in 1474. Near by are the ruins of the sixth- to seventh-century chapel of Sv. Petar.

The islands came under Roman rule in 177 B.C. and were called Pullariae. After the fall of the Western Roman Empire, the rougher Ostrogoths arrived, followed by Byzantium until 778; then the Franks who, under Charlemagne, rampaged across Istria and much of Dalmatia in their merciless christianizing zeal; and finally Venice in 1331. After this the islands were deserted for centuries, because of malaria, and it was not until the nineteenth century that they became popular again. In 1893 one Kupelwieser, an industrialist from Merano, bought them and turned them once more into a luxury resort. He built a racecourse, a golf course, swimming establishments, tennis courts and hotels, as well as cultivating the land in a park and preserving the game. Dr. Robert Koch, the bacteriologist, worked in Brioni for two years on the problem of the malarial mosquito which had wrought such havoc in the past, finally eradicating it completely. Another famous visitor in the past was George Bernard Shaw, and in 1970 Princess Margaret and her husband, Lord Snowdon, were invited there by President Tito. Today many Yugoslavs go there by invitation and stay in one of the two modern hotels, but the islands are closed to the general public.

During the President's time the livestock has been increased considerably. There is a kind of wild sheep, known as *muflon*, with tremendous curlicued horns and tails like deer, while peacocks chatter their way to roost in low trees or on small buildings. Apart from these naturalized creatures, which are so numerous that they appear as if posed for a Constable painting,

[1] *Zadruga* was a collection of families, or a clan living communally and accepting responsibilities, work, etc., under the authority of the most senior — the father, grandfather or even grandmother.

there is a substantial menagerie of animals collected from other parts of the world. Camels, elephants, zebra, emus, antelopes and bears, including three brown ones from Bosnia, live in an ideal zoological garden with a well-equipped veterinary surgery. Tito is also keenly interested in gardening, and the greenhouses and modern propagation equipment in the hothouses on Brioni are reminiscent of a miniature Kew Gardens. Brioni contains a small natural history museum which includes a room of prehistoric remains, such as the actual footprints of dinosaurs, found on the island, besides examples of stone and fossils and stuffed birds and beasts. A second museum houses archaeological finds, including Roman stone fragments and glass. Apart from the animals in the zoo, there are two beautiful Lipicana horses from the famous Yugoslav stud at Lipica in the north of Istria.[1] This gentle pair, so pale in colour as to be almost white, are harnessed to a splendid fiacre which carries one gracefully, with the slight side-swing of an opulent dowager, along the paths to various points of interest.

In character, Brioni is very different from the other Adriatic islands, lacking the ruggedness and the poignancy of most, but possessing a gentle beauty and peacefulness. Perhaps the atmosphere derives from its ancient past, when, in common with the whole of Istria except for a narrow strip down the eastern coast, it belonged to metropolitan Rome and was not an 'outpost of Empire' like the rest of the eastern Adriatic. While the island has always sheltered and protected the seaport of Pula on the mainland, Pula, in turn, has been spiritual and strategic 'big brother' to Brioni through the centuries. From the time of Caesar Augustus, Pula has been a centre of culture and civic affairs. Much earlier it had been an Illyrian settlement: there are the remains of a Bronze Age wall around it. It was mentioned as 'Polai' by the Alexandrian poets Callimachus and Lykophron in the third century B.C. in connection with the legend of Jason and Medea.

[1] This stud farm, established in 1580 in the west of Yugoslavia near Trieste, still has a pure breeding stock of over 100 horses and provided the original stock for the famous Spanish Riding School 'Lipizzaners' in Vienna.

They wrote that the Colchids went to find the lovers and bring them home, but that they could not do so and, too frightened to return without them, formed a new colony called Polae. According to Callimachus, Polae, in the language of the Colchids, meant cities of fugitives. Other theorists claim that Pola or Pula is an Illyrian word meaning 'fresh-water spring'. The nearest map-reference Callimachus and Lykophron gave was the eastern coast of the Adriatic sea, which might refer to any number of habitations; but Mela Pomponius and Pliny both mentioned Pula as a settlement founded by the Colchids. Certainly, the pre-Greek Illyrians were there in strength; they have left not only their defence walls, but also an extensive necropolis, with cremated bodies — the Illyrian practice — of the fifth and fourth centuries B.C.

Until 11 B.C. Pula was probably Illyrian territory dominated by Rome, although nearby Nesactium had fallen to the Roman legions as early as 177 B.C. But Illyrian resistance to Roman expansion continued for several hundred years. After 11 B.C. Augustus moved the Italian border eastwards to the river Raša, and Pula became part of the tenth Italian Region with the name 'Venetia et Histria'. Pula has an abundance of Roman remains which far outshine most other examples. The huge arena, close to the sea, has its elliptical outer walls almost intact and completely unrestored. The amphitheatre measures approximately 132 × 105 metres and is 30 metres high; it had a seating capacity of 23,000. The foundations of three tiers of seats remain, and one tier has been restored to show what the original looked like. Near the main entrance there are the remains of small stone-walled cages or cells. It is thought that some of the wild animals were kept there, to show the crowd what they could expect in the way of a performance — after they had paid for their seats — rather like the side-shows at a modern circus. The huge towers built into the outer oval walls are still standing and were used for various scenic effects — perhaps to simulate rain when needed, and even to perfume the arena. The towers also controlled the canvas covers which shaded the arena; these could be manœuvred like sails and were probably worked by sailors from the port (Pula has always been a great naval base). The arena floor was made of wood and covered with sand to soak up the blood. It was supported on stone pillars which still exist to a height of about seven feet in the subterranean area below today's gravel surface. One can still descend the original stone steps and walk along the passages to the underground halls where animals and condemned humans waited for the performance. At auditorium level there is one passage on the landward side which, it is thought, was used by the gladiators to enter the arena from their college or training-school. Gladiators occupied a special place in the life of the Roman community, just so long as they remained 'on top'; a position similar to that of footballers or bull-fighters today!

Below the arena there is now a small display of Roman artifacts: an olive press with a round stone wheel in working position on a circular grooved stone; and a square wine press operated by a huge wooden screw. The accompanying large, flattish receptacles for the oil and the wine come from a nearby Roman villa. This villa was still in fairly good repair until the end of the Austrian occupation in 1914 – 18, but the front was later destroyed by the Italians to whom Istria had been ceded under the Treaty of Rapallo in 1920. From the beautiful front portion of the building the Italians took the Roman blocks to manufacture slaked lime.

Besides the amphitheatre, Pula has many other well-preserved and beautiful Roman remains. Within the old town is the Forum, still an open square except for one building. At one end is a temple dedicated to the goddess Roma and Augustus, built at the time when emperor-worship came into fashion. This temple, small, tall and quite lovely, has steps, portico, deeply carved friezes and Corinthian columns still intact. Next to it stood a second and similar temple, though now only the back wall can be seen; it was probably dedicated to Diana, or possibly to Hercules. One of the city gates also was dedicated to him, and it has his bearded head carved on the arch.

Other features are the Triumphal Arch of the Sergii, an octagonal mausoleum from the end of the first century B.C. and two Roman theatres. Of these last only fragments are now left of what were, in their time, richly built stone constructions. The larger of the theatres had a frontal façade of about 100 metres. In 1632, during the Venetian occupation, the stone of this theatre was taken to build the fortress at the top of the hill, while the marble columns and portico, it is believed, went to Venice for the building of S. Maria della Salute.

The Venetian fortress is still almost intact and now houses an interesting small war museum which shows graphically just how much Istria suffered before, during and after the Second World War until it was united with Yugoslavia in 1947. Tito finally settled the vexatious question of the Trieste Zones in 1954.

In 1969 the Yugoslav tourist authorities applied to the United Nations Development Programme to extend their Adriatic plans in 1970 as far as Istria and the Italian frontier. Officials in Trieste are reported to have expressed interest to the Yugoslavs in the scheme and the possibility of being associated with it; perhaps providing bilateral assistance. This firm move towards rapprochement followed the successful meetings between Presidents Tito and Saragat in October 1969.

Among Pula's medieval buildings there is a small Franciscan monastery which houses a permanent exhibition of frescoes. Its church is simple and unspoilt by baroque influences with a beautiful gilded wooden altarpiece of Christ, the Virgin, and saints, delicately carved and decorated, from the second half of the fifteenth century. Pula contains many other works of art, including the famous mosaics, and the archaeological museum is particularly rich.

But Pula is not just an antique centre, though a particular overlay of Roman civilization still permeates the town; there are many thriving industries, including shipyards, glassworks, cement factories, textiles, knitwear, furniture, tobacco, etc. It is also a very lively tourist area with a particularly beautiful indented coastline and many local specialities. Istrian national dress is spectacular, and so are the dances, called *balun*. The wines and fish dishes are famous. Oysters and shellfish are a speciality, and there is a festival on 1st May each year, during which the local fishing communities celebrate the *Fešta od Raki* — the catching of succulent round red spider crabs which have many long legs but no pincers. There are yacht marinas and luxury hotels, unspoilt wooded camping sites and tourist 'cottage areas', winter swimming-pools and a constantly mild climate. Best of all, the tourist authorities in the area are well aware of the riches they have to offer, but equally conversant with the dangers of badly planned and overcrowded popularity.

Krk

Istria, with Brioni, is in many ways an area on its own. The stresses and strains of history have been very different in Dalmatia. East of the Istrian Peninsula, the Kvarner Gulf stretches out with the international port of Rijeka at its head, and encloses a whole group of islands. Nearest to the mainland, and to Rijeka, is Krk — pronounced as it stands with all three letters taking equal weight. In fact its name seems to reflect exactly the uncompromising character of this highly individualistic island and people.

Krk is the largest Adriatic island and covers 409 square kilometres. It is close enough to the big world for its inhabitants to commute to the mainland, using the bus and the car-ferry rather like a suburban railway, and yet remains such a unique community that staying there is like visiting a separate country. Various Jadrolinija boats from Rijeka, Zadar and the other islands in the vicinity also call at Krk, while the completion of the new International Airport in 1970 opened the island to the world.

Krk is a land of contrasts: of lush, fertile valleys lying high up in the interior between folds of harsh, bare limestone mountain; of warm green and blue horseshoe bays, sometimes almost landlocked; and of olive groves and vineyards ripening above precipitous cliffs. In the Adriatic, a man from an island, though he may live elsewhere for forty years, remains a man of that island and its way of life. The people of Krk for 400 years enjoyed the protection of the great Krk family, the Frankopans, who controlled the whole Kvarner area. This attitude of overlordship is still apparent today. In springtime the island is particularly lovely, with hellebore and wild cyclamen clustered in great masses in the fields and ditches, and the water cascading off the mountains.

A ride across the island on the early morning bus to Rijeka provides both a good view of the countryside and an insight into the changing life of Yugoslav women. There are the old, perhaps illiterate, but wise peasants in their traditional black; the more worldly middle-aged women with coiffured hair and grey suits; the female executives and professor types; and the young women who are the new office-workers and the tourist guides. The women of Krk island are both pretty and forceful, a characteristic which seems to continue into old age. I shall never forget one elderly grandmother, who clearly did not know the meaning of the word compromise, clambering aboard the bus draped in packages for the vegetable market in Rijeka, about 30 kilometres away. She had tender spring vegetables in brown paper and string, and ancient black plastic bags full of cheese and radishes and, for all I knew, live rabbits. She was pink with exertion and had two teeth missing, but none of her faculties. She held her stout black umbrella upside down, until she used it to bash the inattentive bus conductor on the leg, demanding help with her packages. Catching sight of a distant vacant seat, she then dragged her plastic containers through the legs of another passenger to get there. Arrived, and having knotted her ticket and money safely away, she sat back satisfied with the result of her first skirmish of the day.

The island has five main centres of population: Omišalj, Krk town, Punat, Baška and Vrbnik. There is also a newer holiday resort at Malinska, and one just developing, within 10 kilometres of the new airport, called Njivice. In Roman times Krk town was known as Curicum, and in 49 B.C. the opposing fleets of Pompey and Caesar fought just offshore. Now, the little town is peaceful enough, though it still contains, for me at least, a certain tough awareness. The medieval town was confined within the stone walls and guard towers of the Frankopans' castle, the Prince's residence from 1191. The walls of the keep are four metres thick; a tower stands against the sea and, at its base, there is a sea entrance where boats could land their passengers directly into the shelter of the keep. The main tower is three storeys high, and it is still possible, with care, to climb to the top. One exit at this highest level gives on to the top of the broad wall, forty or fifty feet up. Shakespeare's *Hamlet* has been performed here.

Today the population is 1,320; in the time of the Frankopans it was 11,000 to 13,000, but in Roman times there were even more — 15,000. So it is not surprising to find not one but three ancient churches all huddled together. They are in a narrow paved street which runs uphill from the sea. A shadowy arch spans this passageway, and there are church doors to right and left. The earliest church is the crypt on the left-hand side, built in the tenth century. Simple and low, it has wide and unadorned arches which cross each other to form almost igloo-shaped domes. The biggest of these arched spaces must be 40 by 30 feet across. Concerts of religious music are given here in July and August, and the acoustics are remarkable. The second church, dating from the eleventh century, was built on top of the first, extending across the arch of the roadway, and therefore with three instead of two apses. This Romanesque basilica, Sveti Kvirin, with its decorated façade and murals, is mostly original, though one side of it has been repaired after damage by Allied bombs during the war. On the other side of the arched street is the door of the cathedral. The present building is early twelfth century, but it was originally a Christian basilica of the fifth and sixth centuries, built on the remains of a Roman bath. Krk has been a bishopric since the fourth century, and the Bishop's Palace houses many beautiful paintings. The present cathedral, with nave and two aisles, has been reconstructed and expanded from time to time, the interior columns dating from antique — the Slav term for the heyday of the Roman Empire — to Romanesque, the period when the remains of the Roman Empire lingered on in the Adriatic and particularly on the islands. Among the works of art there is the famous Frankopan silver relief of 'The Madonna in Glory' by the master silversmith Koler, dating from 1477, various Gothic works and carvings, and some paintings, including the polyptych of Sveta Lucia by Paolo Veneziano from the Benedictine Abbey near the town of Baška, a few kilometres away.

Krk is crammed with old stone houses in narrow crooked streets within the original town walls and fortifications. The oldest is the square tower at Kamplin, 1191, but the fifteenth-century walls and bastions near the sea were built by Count Nikola Frankopan. Within this medieval conglomeration one can spot, suddenly, scraps of really beautiful Roman stone carving, not 'on display' but casually let into a wall as decoration, or used as the head of a doorway. The people of Krk have lived with these works of art so long that they are simply things of everyday use and not tourist attractions. To me the most remarkable example of this is a floor of Roman mosaic, of the black-and-white period, now in the living-room of a private house. In fact, the floor actually extends under the main walls to cover rooms in two neighbouring houses as well. Of fine mosaic and delicate design, there is a horned man with a double fish-tail, two dolphins and some sea-serpents in the part which is visible.

On Krk, the season starts a few weeks later than in some resorts — about the second week in May; in March and April, the best time for the wild flowers, the weather can be unpredictable and the islanders still wrapped in their own affairs. Unlike the plushier tourist hotels, Hotel Krk, by the harbour, is always open. It is adequate, spotless, has cold running water, bedroom heating in winter, and a bath on demand.

Punat, the next small town, lies in a wide bay facing south, only ten minutes away. Approaching it, the island bus passes the tiny early Christian church of Sveti Dunat. It lies neglected on a corner, a small, domed hump of rough-hewn stones which stick out of the mortar in points like almonds on tipsy-cake. With a primitive rose window and a rough stone porch stuck on almost haphazard, you have to look twice to realize that this tumble of stone is, in fact, a church.

At 5.15 a.m. the only people to be seen in Punat were a handful of fishermen moving towards their boats. The water inside the lagoon is shallow and, even with a strong south wind, quite calm. When the big ships call at Punat they stay outside the harbour and pick up passengers from small boats. I walked towards the open sea at the entrance of the bay. At the edge of the water the tamarisks were in full bloom. They stood in great billows of blossom the colour of crushed loganberries and cream, pouring out an overpowering scent of honey and heaving, in the strong, hot wind, like women in ecstasy. A few small boats with outboard motors pulled away from their moorings, making for the open sea. Life was stirring for the day.

The land slopes gently from the lagoon. The houses are small, with gardens crammed full at this time of the year with tender seedlings and cuttings. Roses and carnations jostle potatoes and inch-high forests of lettuce. A lean southern cat skitters across one's path, and the sparrows in the oleanders set up a deafening chirrup. On these islands animals are well kept and oxen till the fields, which are lush in the central valleys. The plough is frequently guided by a woman. Life for them has changed little in a thousand years.

Every year, in the early morning of the first Sunday of May, there takes place at Punat a ceremony called *Igra pod Tršnje* ('dance under the cherry tree') — a festival for the young men coming of age. It is a private affair, not a tourist attraction; in fact tourist offices away from the island are curiously unaware of it. Boughs of cherry are fixed high overhead, near the harbour. Long scarves are attached to the branches, and the young men, the *Mladići*, dance round, winding the scarves as they go. It is reminiscent of our ancient Maypole Dance, both probably deriving from a fertility rite, perhaps from the Roman Floralia. After this day the young men are deemed eligible for both marriage and their tour of conscription in the army. In summer, on the second Saturday of August, there is another festival at Punat, the *Puntarska Noć*, which is most fully shared in by tourists. It is a tremendous 'do', with fireworks, decorated 'viking' boats which are specially made in the local shipyard — of traditional, long, steeply curving planks, wooden-pegged — and a figure of Neptune appearing from the sea. This is not to be confused with another splendid get-together held at Krk on the first Sunday in August and called *Ribarski Dan* ('The Fishes Day'), when the fishing nets are thrown out and hauled in by local fishermen and tourists alike. The entire catch is cooked on the beach and eaten in the fingers with maize bread and, of course, buckets of good wine.

Punat was the ancient seat of the bishops of Krk. There are elegant eighteenth-century houses with outside stone staircases to first-floor terraces. Present-day industries include wooden boat-building, figs, vines and olives, and there is an interesting eighteenth-century olive-press. In the middle of the shallow, lagoon-like harbour lies the little, low monastery island of Košljun; about half a square kilometre in area, it is still used exclusively by Franciscans.

Košljun is covered in lush Mediterranean trees and shrubs, interspersed with grape vines, olive groves and patches of vegetables. It has been inhabited since Roman times. A Benedictine monastery was first founded there in the twelfth century, before being taken over in 1447 by the Franciscans who built a new monastery with cloisters. These are beautifully arched in stone, with a stone well in the open centre. There are still some traces left of the older Benedictine foundation; part of the walls and the chapel of St. Bernadine survive.

In summer there are many small boats round the perimeter of the harbour to take one across to Košljun, but in early spring I made the five-minute journey in the monastery boat, a small cabined motorboat with a diesel engine. It had come across with one of the Brothers to do the shopping and collect the post. The Franciscan Brother had sparkling silver teeth and a simple and inquiringly direct gaze. Under his brown habit he wore blue fisherman's trousers, in which he looked very much at home.

In the monastery church, among other notable works, there is a classical painting of the Last Supper on one side wall. On the opposite side is a second painting, only two or three years old, of people in modern dress grouped with some of the monastery Brothers, and a modern Christ teaching the villagers. It is in bright colours, with sea and sky, a good representation of Punat and some recognizable faces among the friars. There is also a movingly simple older chapel. Outside, in a green world, are tree-shaded paths and walks and a small walled courtyard with a chapel. Also outside are thirteen stations of the cross in pottery bas-relief and protected by little roof-hats; the fourteenth station is inside the chapel. Another glade leads to a chapel of St. Francis, as close to nature as the saint could possibly have wished.

The monastery houses a large museum with an overwhelming variety of exhibits. One wanders through an Illyrian and Roman lapidarium, an ethnological department complete with kitchen scenes, cabinets of pottery and examples of national costumes, a library rich with early missals on vellum and early examples of Gregorian chant, and, most astonishing, a zoological section. This is stuffed full of snakes bottled in alcohol, Siamese-twin lambs, *škrpinj* (a fearsome-looking fish but one of the best to eat), sea-horses, small sharks, jackals, foxes, eagles and lizards, and one small stuffed lamb with only one large eye placed in the centre of the head like a Cyclops. 'Ah, these things happen with animals as well as humans,' murmurs the Brother.

The coast road peters out at Punat, but inland it crosses the island almost due north to arrive at Vrbnik (which faces the mainland) or, on an easterly branch, runs across high ground to the valley which ends at sea level in the small town of Baška. The latter has a population of about 850, but has become a popular summer tourist resort. The area, with its richly fertile valley of Baščanska Draga, was inhabited in prehistoric times. Later, the Romans developed a major settlement there, close to the sea. Numerous remains, architectural foundations, mosaics and tombs have been found, particularly near the Romanesque church of Sv. Marko. The ruins of a medieval village burnt by the Venetians in 1380 can still be seen. A new village then developed close to the sea, with a long fine shingle beach and some delightful small sheltered coves.

About a kilometre inland is the village of Jurandvor. Here is one of the keys to early Croatian culture. On a piece of land given by King Zvonimir, who was crowned King of Croatia by Pope Gregory VIIth's legate in 1076, Benedictine monks built an Early Romanesque church dedicated to Sv. Lucija, which was completed around 1100. It was at this church that the famous Baščanska Ploča was discovered. The inscription on the stone is carved in the famous Glagolitic lettering and records Zvonimir's gift and the building of the church. There is now a good copy of the stone in the church, but the original is in the museum in Zagreb. The polyptych of St. Lucy by Paolo Veneziano was the altarpiece, but this too has been removed and is now in the Bishop's Palace in Krk town. It is strange to think that this almost forgotten and ancient little church held a tangible piece of history so important to the proving of an independent Croatian culture during the early Middle Ages.

Baška faces the Velebit Range of mountains on the mainland and is at the eastern end of Krk. Due south lie the western extremities of the island of Rab. Nearer to hand are the small islands of Prvić, Grgur and Goli, uninhabited except for a detention centre.

Facing north on the island of Krk is one of the most exciting small, defensive medieval towns: Vrbnik. Vrbnik still, in the twentieth century, glowers across the water from its cliff-top perch towards Novi Vinodolski and, farther east, Senj, the two mainland ports which were so tied up with the powerful history of Krk. Senj, with its square, uncompromising castle on the hill, came under the control of the Frankopan counts in 1271, acquiring a statute in 1388. It was not until 1469 that the Frankopans lost the stronghold when it was taken out of their control by the Hungaro-Croatian king Matija Korvin. Later, Senj and the castle were the base from which the *Uskoci* — refugees from the Turkish-controlled interior — engaged in a bitter struggle with both Turks and Venetians. They fought on land and on sea, leading the local people; they caused war between Venice and Austria in 1615, and were instrumental in the great build-up towards the Battle of Lepanto and the final Turkish naval defeat in 1571. An élite band of pirate/patriots, they watched the sea and turned its vagaries to their advantage. The strip of sea between Krk and Senj, which wears a silken look in summer, can rage with the ferocity of a tiger in winter when the *Bura*[1] blows. The unsuspecting Venetians would get into difficulties, and the *Uskoci* would weigh in and carry off the prizes. When not at sea, they attacked Turkish strongholds on land. Many ballads extol their pirate virtues which, in some ways, were a prototype of partisan warfare. The medieval back streets of Senj still have a swashbuckling atmosphere thick enough to cut with a knife and should not be missed by anyone staying on Krk, which has a direct ferry service from Baška to Senj.

The second mainland connection with Krk, which has an occasional passenger boat service to Baška as well as the normal car ferries, is Novi Vinodolski. The centre of this small town is the tough thirteenth-century Frankopan castle. Here the Vinodol Code was drawn up; an important landmark in early Croatian common law, written in 1228 in the Glagolitic script. The castle itself was so renovated in the nineteenth century that the original is hardly recognizable; but one has only to look across the water from Vrbnik, a maze of crooked streets and hidden doorways, on an almost vertical cliff 50 metres above sea level, to remember the intrigues of those stirring times.

[1] North-east wind.

Vrbnik now has a population of about 1,200. It produces wines, vegetables and tomatoes. There is a steep road down to the small harbour. Apart from a few prehistoric remains — graves have been found in Sveti Marak cove — the first records of the town date from about 1100. It was a fortified town of the princes of Krk and its statute was drawn up in 1388. There are remains of the defence walls, towers, a fort, two Gothic chapels, and fifteenth- and sixteenth-century works of art. But the most interesting exhibits are in the Zadružni Dom, or municipal building, which houses the library of a politician, Dinko Vitezić, who was born in Vrbnik. There are about 15,000 volumes, including a copy of the *Atlas Scholasticus et Itinerarius* (1718), some old illuminated missals of Vrbnik — one of 1463 — and some Glagolitic documents and several incunabula.

It is hard to realize that Krk and its towns were once so affluent and so powerful, though the stamp of proud independence is still visible in the features of the islanders, as well as in the richness of their national costumes and medieval headdresses, dances and songs. Krk was governed, in spite of the general overlordship of bigger powers such as Venice and Croatia, by the Frankopans, one of the most powerful families in Croatia, from the twelfth to the end of the fifteenth century. During that time they, and their friends the powerful Zrinski family who had a castle at Bakar on the mainland, wielded a power whose influence was felt far beyond the island to Zagreb and other centres of European friction and diplomatic struggle. Their lands spread across the mainland from Trsat, on the hills behind Rijeka, to Senj. Even after they had left their island of Krk to the care of the Venetians in 1480, the name crops up again in the seventeenth century. The last members of the two famous families, Frankopan and Zrinski, made an unsuccessful attempt to free Croatia from the domination of Austria. Ban Petar Zrinski and Prince Krsto Frankopan were taken to Vienna in 1671 under false pretences, sentenced as traitors and beheaded. With their failure, the last chance of successful resistance to Austrian centralization and Hungarian domination was lost; it did not really revive until the cultural and literary stirrings of the nineteenth century.

Nevertheless, island customs have always stayed very much alive, and Krk is rich in folklore and tradition.

Omišalj, which now has modern hotels and a tourist resort close to the sea, is a medieval fishing village perched 85 metres above the shore with superb views across the Kvarner Bay. The old houses are grouped round what was yet another stronghold of the Frankopans. The parish church, recorded in 1213, has fragments of ninth- and tenth-century ornamentation, and Glagolitic inscriptions. In a small bay close to Omišalj have been found the remains, including mosaic floors, of what was probably the Roman town of Fulfinium. There are also the ruins of a fifth-century basilica. On 15th August each year the 800 inhabitants celebrate *Stomarina* ('first fruits'), dancing and singing under festoons of silk scarves and fruit in their beautiful costumes, rich with lace, silk-fringed shawls, and coral and other necklaces.

Old and new blend well on Krk. Malinska, grouped round a wide bay, has developed only in this century, yet the buildings melt into the natural scenery so well that there is no sense of anachronism. Krk has not the feeling of remoteness of some of the other islands, perhaps because it always busied itself with the Big World during its medieval heyday. Although its island affairs are small these days, one gets the impression that it still holds a watching brief over the large mainland area it once dominated.

The castle of Trsat, above Rijeka, for example, built before 1288 and recorded in the Vinodol Code, was Frankopan; it remained with the family until 1671. Near by, besides the thirteenth-century parish church of St. George, there is the Frankopan church of Marija Lauretanska — Our Lady of Loretto — which, together with the Franciscan monastery, was built by the Frankopani princes in 1453. The church has long been associated with miracles, and so many pictures were brought as thank-offerings that they are now kept in a seperate, modern, but equally attractive, new church close by.

The castle of Trsat was taken over by an Austrian vice-marshal, Laval-Nugent, in 1826, when it was restored and enlarged. He included a mausoleum, and it is of interest that the aunt of George Bernard Shaw was buried at Trsat, and her body has only recently been removed to Rijeka. The gardens now contain a delightful open-air restaurant and stage, providing a pleasant evening's excursion to the heights above Rijeka. There are frequent buses, and there are also 400 steps which hairpin down into the town and are used by pilgrims to the monastery church.

Rijeka

Rijeka, too, is a good deal more fascinating and historically rewarding than it might at first appear. The old town lies behind the modern international port, and there are some buildings, through the Old Gate and under the Bell Tower (1377), which, like its cathedral, are well worth exploring. The museums have many interesting exhibits from the neighbouring islands, and there is now a new Glagolitic[1] Museum, opened in 1969. The Glagolitic script and the Cyrillic version of this Slavonic alphabet (which appeared a little later, 860 – 70, and was introduced by the Brothers Cyril and Methodius) marched side by side as forms of writing suited to the Slav language, instead of Latin. Cyrillics were used — and still are — in Serbia, Montenegro and Macedonia, while Glagolitics, evolving from 'round' to 'square', were used in Croatia and in Dalmatia which is part of Croatia and elsewhere. The Glagolitic text became a banner under which the churches fought to celebrate their services in their own language, written in their own text, instead of in the Latin of Rome. It also became the vehicle for Croatian literary culture. Perhaps most important to the South Slavs is the fact that Glagolitic inscriptions on stone, whose veracity can nowadays be tested by scientific means, are proof that their lands were, in the Middle Ages, Croatian and not Italian.

[1] The alphabet first evolved to write down the Croatian language.

Chapter 2 The islands of Cres and Lošinj, Ilovik and Susak

Cres and Lošinj, the main islands in the outer Kvarner group, are usually coupled together in people's minds; almost like twins. Indeed, they are actually joined at their narrowest point where Lošinj nudges Cres from the west. Both islands are very long and very thin, causing a considerable barrier between the eastern and western halves of the Kvarner Gulf; but the Romans dug a canal across the few metres of land which join them in order to get their galleys through. The Romans were notorious for never taking to the open sea if they could avoid it, and the canal saved them hours of sailing on their journeys between Rome, Istria and the south-eastern Mediterranean. The Roman canal is still useful for small craft.

But though the islands of Cres and Lošinj are joined, and so could be regarded as one, so different are they in character and atmosphere that they are as unlike as any pair could be. Lošinj is extrovert, while Cres is withdrawn; yet both are a complete change from the tough strength and forceful character of Krk.

There is a regular daily boat and a car ferry, both run by Jadrolinija from Rijeka, and sometimes a fast hydrofoil service to Lošinj; but for those who like time to absorb the atmosphere, even on the outward journey, the passenger/cargo boat is the most rewarding form of travel. Disgorging passengers and cargo on its rounds, the boat wends its way from port to port, calling in at Cres and some smaller islands, until it finally comes to rest, way out from the mainland, in the long sheltered inlet which forms the huge natural harbour of Mali Lošinj. At Porozina, the most northerly port on Cres, the ship drops a number of her regular passengers, who then take a bus across the island along the 85 kilometres of asphalt road. This runs the whole length of both Cres and Lošinj from Porozina to Veli Lošinj. Sometimes the boat goes on to call at Unije, a small island of 18 square kilometres off the north-west coast of Lošinj. The only village on Unije has about 270 inhabitants and is like a fishing village on one of the Scottish western isles. The port is a wide, low, natural scoop, and the village rises steeply up a hill as bare as the western Highlands. Seagulls keep station outside the harbour, floating on the water like ornithological traffic cops. The red-roofed stone houses stand square against the wind — which can blow in winter! The villagers crowd cheerfully on to the quay to see the boat and their friends arrive. Life is uncomplicated by clock-time, the seasons still count, and the important things of life are the quality of the year's wine and olive oil, and the last catch of fish. For anyone seeking isolation and peace, Unije is as good as any Hebridean island, with the advantage of guaranteed blue skies and hot sun during the long summer. There are 195 tourist beds in the island, all in private houses.

The boat from Rijeka continues to Mali Lošinj, taking between five and six hours according to the number of stops on the way. A newcomer may be surprised by its calling at Mali (small) Lošinj rather than the other town, Veli (great) Lošinj, both of which have harbours! But these prosaic ideas soon pass as the visitor gets swept irresistibly into the inconsequential gaiety of the island. Great or small Lošinj, what does it matter? In fact, even the names are wrong, for Mali Lošinj has a bigger population than Veli. The island has an atmosphere like the naughty days of pink champagne and glass slippers. It floats, like a spare and giddy rib, out to sea from its quieter sister, Cres. It not only floats, but it swings! It is swinging summer and winter with groovy 'luxe' hotels, with sizzling air and a sparkling sea, and an air of everyday carnival round its port.

What makes Lošinj such a vivacious and feminine place? What gives it this air of easy laughter and good nature? Part of it is surely attributable to its climate. Although this is variable in spring and autumn, the rain never lasts very long, and the atmosphere gives a rare purity to the sun's rays. The island is far enough away from the land mass of southern Europe to be comparatively unaffected by the mountains of the Julian Alps and the Velebit range. It is also so slim that in places it could almost be compared to an outsize ocean liner, enjoying sea at close quarters on all sides. There are trees of many species on Lošinj, brought home by seafaring men from all parts of the world. The profusion of these trees and the clear atmosphere seem to affect even the birds, which sing in odd keys: the humble pigeons are apt to sound drunken and baritone, and get their mating calls all wrong!

The town of Mali Lošinj has a long and full seafaring tradition, with a nautical school and a shipyard, although, like so many others with a wooden ship tradition, bigger yards and modern methods have taken away much of the old trade. The houses of Mali Lošinj are of a higher standard than one might expect, the reason being that they were built by sailors who brought home all the ideas they gathered in other parts of the world. The oldest section of the town rises in tiers above the harbour, while the hotel area is centred not far away on the wooded slopes of the Čikat cove. The tourist industry has been thriving since the Austrian emperor, Franz Joseph, built his 'Villa Karolina' there, towards the end of the nineteenth century. It is a fantastic baroque structure in one of the best positions, shrouded by pine trees and close to the sea, with elaborate baroque landing-stages. It is now a luxury hotel. It is an interesting point that, in those early days of tourism, it was the winter season that counted, just as it did in Monaco. Some of the hotels now have special group-price amenities in the winter months for older people who have heart or asthmatic conditions necessitating a good climate. The island also has special therapeutic clinics. At the other end of the scale, on 31st December and 1st January, the whole area hums with the International Underwater Fishing competition, an annual event since 1960.

On one arm of Čikat Bay a small, thick-walled church stands close to the shore. Inside, the walls are hung with pictures, thank-offerings of sailors who have come safely to harbour from perils on the sea. This form of expressing gratitude in seafaring men can be found in a number of areas on the Dalmatian coast, and serves as a constant reminder that the Yugoslavs are a maritime nation with similar traditions of service and the same basic attitude towards the sea as we have in Britain.

All through the nineteenth century Lošinj men were developing their sea trade. Six shipyards were founded, which built ocean-going sailing ships. In 1870 the fleet numbered 131 with a total of 62,000 tons. The island at that time was under Austro-Hungarian rule, and the first sea-captain to navigate round the Cape of Good Hope under the Austrian flag was a Slav from Veli Lošinj; he was decorated with a gold medal, and his name was a good Slav one — Anton G. Busanić!

It is sometimes a little difficult now to realize that a people so strongly Slav and freedom-loving, struggling through all the centuries of their existence for self-expression, should have been so completely dominated by other nations. One reason, undoubtedly, was the smallness of their groups before the emergence of the Yugoslav nation. How could Croatia, alone, stand up to Austria with its Hungarian allies, or Italy, for that matter? Previously, the groups were even smaller, sometimes only half an island, or an autonomous city and its surrounding villages. Towards the end of Austrian rule, however, the struggle for national right of expression began to develop in various parts of Croatia, including the islands of Lošinj and Cres. Here there were two societies, the 'Cyril and Metod' — named after the monks who had created the Cyrillic alphabet — and the 'Zora' society. With the help of these two bodies in the islands the first Croatian language schools and a reading room were opened in Lošinj.

Mali Lošinj was first recorded as Malo Selo (Little Village) in 1398. The church of Sveti Martin is fifteenth century, and so are the Venetian observation tower and the nearby church. But the islands of Lošinj and Cres were inhabited much earlier. There is a mass of evidence to show that the area was considerably populated in prehistoric times, when the original Mediterranean aborigines of the Neolithic, Bronze and Iron Ages made good use of caves and grottoes in the southern part of the islands. Bronze Age and Iron Age Illyrians built strongholds on the peaks and fortified them with dry-stone walls. Crypts and crematoria, ceramics and amber ornaments, have been found, particularly near Osor, showing the latter as an important historical landmark on the ancient Amber

Route. In the third century B.C. Apollonius of Rhodes wrote about Apsyrtides in his great legend of the Argonauts and their exploits in the mythical period of about 1200 B.C., and he mentioned Absyrt, the ancient name for Osor. Not much has been found of an ancient Greek settlement, but according to legend the Greeks did arrive, only to leave again after two years of drought. The Romans, on the other hand, left numerous remains in the form of country villa foundations; it was a fashionable place for them.

All the same the Illyrians, particularly in the interior, continued until A.D. 6 or 7 to harass the conquerors; an early form of the partisan fighting which became famous in the Second World War. In remote areas, there were pockets of Illyrians almost uninfluenced by the foreign invaders until the coming of the Slavs.

Like Mali Lošinj, the sister town of Veli Lošinj was first mentioned in 1398, its name then being Velo Selo; its oldest church is the fifteenth-century Sv. Nikola. There are some sixteenth-century Venetian fortifications, built to protect the town from attack by the Uskok pirate-patriots of Senj. Veli Lošinj is very close to Mali, but on the other side of the narrow island, facing east. It has two harbours, neither of which is very well protected during bad weather; the second, the attractive little fishing port of Rovenska, is the safer of the two for yachts. The population of Veli Lošinj is about 850, whereas Mali has nearly 4,000 inhabitants. The two towns had a similar history until the Romans finally withdrew. During the Middle Ages both seem to have been uninhabited except for a few hermits. It was not until the Slavs were well established on the mainland that they left evidence of their arrival on Lošinj and Cres. The first Slavs to move in were recorded in 1280, very late compared with other parts of the Dalmatian coast. Their settlements were granted autonomy by the feudal overlords of Osor in a contract signed in 1389. It was these new Slav people who built the first church, Sv. Nikola, at Lošinj in an area named after the first Croat family to arrive, Hargović. Churches played an important part in the everyday life of the people, usually occupying the centre of local culture and learning. The priests were also public notaries and used the Glagolitic characters for all documents. After 1442, until the early nineteenth century, the Croatian priest on Lošinj used the Glagolitic church services, defying the edicts of the Vatican.

The five ecclesiastical buildings of Veli contain some interesting paintings and other works of art. To the church of Sv. Antun one sea-captain brought five marble altars from Santa Croce in Venice, as well as various paintings by Venetian masters; the most famous of these latter is the 'Madonna and Saints' on wood by the fifteenth-century Venetian, Bartolomeo Vivarini. The stone tower in the town is genuine fifteenth century, but the impressive castle, built just above the town, is of much later date. It was erected in 1885 by the Archduke Karl Stefan of Austria, who also planted a remarkably luxuriant park of trees and shrubs.

From a historical point of view, perhaps the most fascinating of Lošinj's treasures is a picture, now in the church at Veli, brought from the small offshore monastery island of Palaciol. It is a beautiful Orthodox icon, painted on wood, of the Virgin and Child with, underneath, five saintly figures. Four of them are in Orthodox robes in attitudes of benediction while the fifth, bearing a lance, is on horseback. How did this icon of the Orthodox Church find its way to Lošinj when all Croatia has always been Roman Catholic? The reason is that many Greek priests and monks escaped from the infidels who overran the eastern monasteries at the time of the Crusades, and a handful of Brothers travelled the length of the Adriatic to settle on two very small islets, Palaciol and Oruda, which lie six or seven kilometres out to sea from Veli Lošinj and Rovenska. The priests brought with them some of their treasures. They built their church on Palaciol in 1206, and their monastery on the twin island of Oruda, and they provided the monastery with good underground cellars in which they were able to hide their treasures.

It is delightful, on a calm summer's day, to take a small boat from Rovenska, a little fishing community, and wander along the coast dodging in and out of the rocky nooks and crannies. The sea is very blue and the rocks form arches overhead at the entrance to watery grottoes. At the same time one can slip across for a look at Palaciol and Oruda.

From Mali Lošinj the road across the islands takes in many points of interest, including the village of Čunski where they breed sheep which yield astrakhan sheepskins. It passes first along the inner arm of the big, sheltered harbour. Near the northern end of the island the small village of Nerezine lies in a wide bay, completely sheltered on all sides. Nerezine, at the foot of Mount Osoršćica (588 metres), is surrounded by vineyards and olives, has sea and mountain air at hand, sandy beaches, and is drawing more tourists each year. There are hotels and rest homes and some interesting remains of the Venetian past. Each year they hold a traditional Nerezine 'Night', as well as a regatta.

The land as it runs towards Cres is so narrow at one point that there is a man-made canal of a few metres joining the sea from east and west, with a swing bridge to alternate road and water traffic. This canal, by the way, is not to be confused with the Roman one which separated the two islands nearly two thousand years ago.

The Roman canal is at Osor, the ancient Absyrt, across the narrow neck of land where Lošinj and Cres meet. Osor is undoubtedly one of the oldest centres of civilization in the Adriatic. Though it is a modern, inhabited village with children playing around its houses, it has that special 'feel' about it which sets it apart — the air of having watched, in the past, things unspeakable, and times of triumph; acts of grandeur and acts of failure, and, having watched, survived. Osor was mentioned by the tenth-century Byzantine emperor Constantine Porphyrogenitus, who called it one of the most important towns in his empire. Until the fifteenth century it was the main urban centre of both islands; in fact the whole Cres-Lošinj group of islands was called after it — Osorski Otoci.

The Romans came in strength to Osor — to them, Absorus — developing it into a major fortified town. It is estimated that there were at times 40,000 Roman soldiers, though many of them died from malaria. The small museum is full of fascinating and assorted remains. Human bones and tear-bottles have been excavated from graves; quantities of small, flat terracotta lamps with decoration of animals, or erotic scenes like the murals in the house of the Vettii in Pompeii; skulls, one of them Neolithic; amphorae and other vessels; a small rectangular stone box with a lid, in all about 1 foot square, which still contains the cremated ashes, glass and jewellery from an Illyrian grave; and quantities of Bronze and Stone Age artifacts. When I was there the two storeys of the old stone house, which was the Town Hall in medieval times, were being decorated, and its contents were waiting to be rearranged. Everything was piled in confusion in heaps on the floor: bones waited for identification, and small, flat pottery lamps lay about in the sand. Somehow this disarray added to the poignancy and honesty of these relics from Osor's past.

The church at Osor — a cathedral dating from the fifteenth century with early Renaissance architecture — shows the influence of one of Dalmatia's finest stonemasons, George of Dalmatia. The town was a bishopric from the sixth century. During the last war the cathedral was bombed and damaged, but repairs were completed in 1969. The priest willingly shows visitors the church treasure and a Titian picture over the high altar.

A walk up the street past the cathedral brings you to the old grey city walls. There are the foundations of a temple, public buildings, mosaics and the remains of Roman country villas and inscriptions. At the top of the incline the ruin of a Franciscan monastery with its Gothic chapel (1414) stands out overlooking the sea at Bijar Cove. A little beyond the ridge on which it stands, the ground falls to sea-level again, and the Adriatic encroaches almost to joining-point across the flat, narrow neck of land. Some people think that the original Roman canal was dug at this point and not where the present canal and swing bridge are situated at the other side of the town. Osor had a lively medieval history. In 841 the Saracens came and burned down the town. In the next century it became part of the Croatian kingdom, recognizing the sovereignty of the rulers whose dynasty began when Tomislav (one of the župans, or clan-like chieftains, of Nin, a town further south along the coast) assumed the title of King. Tomislav was crowned about A.D. 924, and from then on Croatian history becomes extremely complicated by the cross-currents of Venice, Hungary and Bosnia. In the year 1000 the Venetians arrived and assumed control of Osor. Then, in 1105, Koloman, a Hungaro-Croatian, who had been crowned King of Croatia and Dalmatia in 1102, took control. This event followed the confusion after the death of the purely Croat king Zvonimir in 1089, and the intervention of King Ladislas I of Hungary. Ladislas made claim to the Croatian throne by virtue of being related to the royal house of Croatia; he ruled from 1091 until 1095. Finally, in 1409, Osor — and the Osorski Otoci — became part of the great Venetian empire. The Venetian overlordship was replaced in 1797 by Austro-Hungarian rule, which lasted, except for the brief Napoleonoic French period of 1806 – 13, until the end of the First World War. Even then, though the kingdom of Yugoslavia was formed in 1918, the Osorski Otoci — Cres, Lošinj, Susak and the smaller islands — were not free of their long-endured foreign yoke. Under the Treaty of Rapallo they went, as part of Istria, to Italy, together with the watershed of the Julian Alps, including the important international port of Rijeka (Italian Fiume), Zadar (Italian Zara) and Lastovo (Italian Lagosta). All, in this arbitrary package-deal, suffered from the growing viciousness of Italian Fascism.

At Osor, almost without realizing it, the traveller crosses the canal to the island of Cres. But very soon the atmosphere changes, and the essential bare loneliness of the Cres hills — windswept in winter and with a rugged, stunted, and leaning tree growth apparent in the high, exposed areas — replaces the softer aspect of Lošinj. The island of Cres has a withdrawn, steadfast quality about it which is strangely appealing. The small cultivated fields are overshadowed by bare, grey rocky ridges. On the heights the bee-hives are battened down with rocks against the wind. The valleys once had rich woods of beech trees, but during Venetian times these were cut down to make oars for Venetian galleys, leaving the soil unprotected and diminished. Now there is an afforestation programme, and patches of young trees are springing up once more.

Cres is quite large: 404 square kilometres compared with Lošinj's 74·68. Its most remarkable feature, although there are no surface streams, is a huge freshwater lake called Vransko Jezero, or Lake Vrana. This is not to be confused with the other Vransko Jezero near Zadar, on the mainland. Most of the villages are on the western side of the island, though there are some very interesting and isolated hamlets in the hilly interior, and there is a Neolithic cave-dwelling to the east at Punta Križa. High up on the less accessible heights one can find numerous Bronze Age earthworks surrounded by circular stone walls. Rectangular fortified earthworks from the Iron Age — at Beli, Lubenice and Ustrine — and tumuli still exist. Illyrian Liburnians, Greeks and Romans followed each other. Later, the Slavs from beyond the Carpathians settled on Cres, where the oldest of all South Slav Glagolitic inscriptions was found, the famous Valun Stone dating from the eleventh century. Valun, facing north on the west coast of Cres in a wide bay, has only a lane connecting it with the road. It is an enchanting small fishing port with thirty houses and a population of about 150. The Glagolitic Stone is still kept in the parish church of Sv. Marko. Chiselled in the *obla* or old Glagolitic, not the later square formation, it is probably a gravestone. Surrounded by olives, figs and vineyards, Valun has stretches of fine gravel beaches. There are a few private rooms available for visitors and good food including cheese, *pršut* (the delicious Dalmatian smoked ham), fish straight from the sea and the excellent Lubenica wine.

On the main road, inland and north of Osor, one of the first villages is Belej. Apart from normal hill-village agriculture, Belej has an unusual industry, the rearing of pheasants. On the grassy slopes of fields sprinkled with trees there are cages and runs holding an average of 600 poults, including about seventy golden pheasants. These are not only sent to shoots in various parts of Yugoslavia but are also exported to Italy. Although golden pheasants are no good as game stock, a good price is paid for them in Italy as garden and courtyard pets. North-west of Belej is the small fishing port and sheep-raising community of Martinšćica, not easily accessible from inland. But the boat from Rijeka calls there, and, with 120 private rooms available, the village should be added to the list of anyone who wants a small quiet hideaway from the modern world. Tourism there, by the way, is about 2,000 years old: wealthy Romans favoured it as a holiday resort.

The next village on the central road is Vrana, and very soon one gets a first glimpse of the beautiful lake which is nearly two kilometres wide by four long. It lies in a great bowl hundreds of metres below the road on the left-hand side, a shimmering, almost metallic-looking stretch of fresh water. It is very deep, its bottom being 68 metres below sea-level while its surface is 16 metres above. Its capacity of about 200,000,000 cubic metres supplies the entire population of both Cres and Lošinj; the water probably reaches the lake through undersea channels from the huge Velebit mountains which skirt the mainland coast between Rijeka and Zadar. The land sweeps down to the great sheet of water in grass-covered folds and pockets of trees, reminiscent of the English Lake District. The trees are for the most part cherries, whose blossom in early spring is a lovely sight.

North of the lake, still on the western side of the island, lies Cres town, with a population of 1,900 living round its harbour. Cres was known as Crepsa to the Romans, and there is plenty of evidence that it was inhabited in prehistoric times. There was a Christian settlement with a basilica as early as the sixth century and during the Middle Ages the town developed within some defence walls and received a statute in 1332. From 1459 it replaced Osor as the centre of administration under Venetian rule. Near the sheltered harbour there are plenty of attractive public buildings and private houses from the sixteenth and seventeenth centuries, among them the City Gate, the granary, the clock tower and the loggia where the town Elders used to meet. There is no museum, but one palace, 'Arsan', is being preserved to show how such a house looked in Venetian times. Just outside the town, by the sea, is a Gothic chapel of St. Mary Magdalene dating from 1402, and a Franciscan monastery with two cloisters, in the library of which there are a number of fifteenth-century illuminated manuscripts. In the church, which is fourteenth to fifteenth century, the Gothic stonework and carved choir-stalls are worth looking at; and not far away is a fifteenth-century Benedictine convent. Cres town has a fish-canning factory, a knitwear factory, and a small shipyard which builds wooden craft. Both islands export a pyrethrum dust insecticide obtained from the marguerite flowers which grow in great numbers. They also make a resin from maquis shrubs, which is used to preserve fishing-nets. Cres town, during the Venetian domination, had to pay tribute to her overlords. This involved not only the cutting down of too much valuable timber to make Venetian galleys but other tithes as well. One tax was a certain number of overcoats from the town each year, and rabbit skins to line the rich garments of Venetian noblemen, as well as beechwood oars.

In this part of Dalmatia the dialect varies not only from one island to another but even from village to village. In Cres town they speak a very distinctive form of Croatian: for instance, *mačka*, a cat, is *meska*; *ja sam bio*, I was, is *ja sam bil*; *kede si bil?*, where have you been? is *gde si bio?* On the other hand they use many words without the insertion of the Croat *j*.

Although most of the development of the island has taken place close to the sea, one or two of the inland villages have a unique quality of their own, and a remarkable spirit of independence. Such a village is Orlec, which must be approached over a rough and stony track. I asked if the island of Cres had a national dress different from that of Lošinj, and was told that costumes differed in each village. We went to Orlec to see the most outstanding.

It had been raining, so the small hamlet wore an even more withdrawn air than usual, rather like a convent whose nuns are not sure if they are allowed to speak or not. One or two children peered from doorways, and then a more forthcoming sixteen-year-old recognized our driver. Two or three motherly looking women appeared in everyday peasant black; but when questioned about the local costume they turned away, shrugging their shoulders, telling us to come back tomorrow or saying that they had grown too fat for the bodices. At last one elderly woman offered to show us her costume piece by piece: though she would not put it on we had won the initial battle of wills! She bustled upstairs to collect all the items of clothing and jewellery which had been handed down through generations and worn at successive weddings, christenings and festival Sundays. In the meantime her twenty-year-old son came in and started to speak English. He was a seaman on British merchant ships, and the miracle of communication in a strange language broke all his mother's bonds of restraint. Like sunshine after rain she lavished food on us while we drank the Scotch whisky he had brought home. She dressed my companion in her island costume, layer on embroidered layer, topping it all with a green headscarf tied and pleated exactly to tradition with a good 'pinch' at centre front to form a peak. Lastly came the jewellery: the coral and coloured glass beads and the gold earrings whose pattern has changed very little since the early Slav prototypes arrived in the area about ten centuries ago. There are plenty of similar examples in the museums. Every family in the village owns at least one of these costumes which are preserved carefully in chests, mended and restored when necessary, and brought out for all 'occasions'. By this time the village, on its 'most immediate' grapevine, knew who we were; we were drawn into a second and then a third cottage, plied with sheep's cheese, hot new bread, and *Orahovica*, a grape *rakija*[1] flavoured with walnuts, and eventually found it difficult to tear ourselves away.

Ilovik

Besides fourteen small islets and thirty-one rocks and reefs, there are two other small islands in this outer group of the Kvarner Gulf. One is Ilovik, about one nautical mile south of the tip of Lošinj, with a population of 350. It is hilly and almost all of it is under maquis scrub. There is some sheep-breeding and, of course, fishing. Its harbour is a narrow channel between Ilovik and an uninhabited islet called Sv. Petar off its north-eastern coast which protects the village from the *Bura*, the most dreaded of all Adriatic winds. This can sink a big ship close to land if it catches her on the port or starboard quarter, its great danger being that it does not blow constantly but in great downward gusts off the mountains.

Ilovik boasts prehistoric earthworks and the ruins of a medieval chapel. Its companion, Sv. Petar, has traces of a Roman *villa rustica*, and a few walls from an eleventh-century Benedictine monastery, Sanctus Petrus de Nembis. There is also a Venetian fort built in 1597 as defence against the Uskoks. Ilovik has a small tourist industry with 135 private rooms to let. Idyllic, gentle and inexpensive, it is connected two or three times a week by small passenger ships from Mali Lošinj.

Susak

The second of these small islands is Susak. It has a reputation, based on its complete originality, which far exceeds its geo-

[1] *Rakija* is a colourless spirit, like *eau de vie*.

graphical size. Susak, quite unlike the other islands, is built of sand, and has neither woods nor stones. It is 6·3 square kilometres in area and the highest sandy point or dune is 98 metres. The sand, it is believed, was probably deposited by sea currents over thousands of years and is composed of particles of freshwater crustacean shell. A second theory is that it originated through volcanic eruption from the sea-bed, which would account for the rock which rings the island at sea-level. It would be easy to imagine Susak as a barren spot, but because the sand is of a sort to hold moisture and trace elements it is one of the finest areas for viticulture and is almost covered with terraced vineyards. The view of the island from the sea as one enters the only port is rather like a green and yellow layer-cake with steep sides and squared corners. The old part of the village lies on the steep hill behind the port. Near the harbour are some large fish-halls and, near by, the paraphernalia of oil-drilling. Susak has natural gas, and the authorities have been probing for accompanying oil, but at the end of 1969 this ceased as the drilling depth became too great.

Susak and its inhabitants have always been separate from other communities, though the island is only six nautical miles from Lošinj. In the churchyard there are only seven or eight family names and, because of its isolation from outside influence, the islanders have preserved a very ancient form of dialect, together with a brilliant island dress, practical shoes made of canvas — particularly suited to the sandy paths — and marriage customs, folk dances and songs quite unlike any others. Like the people of Lastovo, further south, they also have a way of finding an appropriate nickname for any newcomer to the island. I have yet to discover what they called me, but I am assured that there will, without doubt, be a name.

The costumes are more brilliantly coloured than any I have seen, and they are worn frequently. Basically brilliant pink, yellow and green, with white, the women's clothes are remarkable for the number of lace-edged petticoats and the shortness of their skirts, many inches above the knee these are undoubtedly the original forerunner of the mini-skirt. The aprons worn over the skirts are vertically striped with yellow, green, bright pink and blue ribbon stitched down to join the frothy and thickly decorated border which swings out almost horizontally above the layers of petticoats. The blouse, with wide, richly embroidered sleeves, is made of a light but thick homespun linen. A brightly coloured shawl with fringed edges goes round the neck and crosses at the waist. The whole is worn with petunia-pink hand-knitted stockings and chrome-yellow canvas shoes. The hair, too, is specially dressed in the Susak manner in ringlets to each side of the temples. For her wedding-day a girl has a gorgeous headdress and, once she is married, wears a dark head-scarf with her island costume. The songs are very curious; sung always to more or less the same tune, with a nasal intonation. Girls perform the dances — an unusual feature in many peasant communities — and the many pirouettes and twirls swing the skirts out like a ballet-dancer's tutu.

Industry is at a very low ebb, apart from production of some excellent wines, which include an aromatic and strong white, resembling a Greek *retsina*, and the famous red *Pleskunac*. Susak's chief export during the present century has been human beings, who have emigrated to many parts of the world. The fish-halls near the harbour, though vast, had only two people working there when I visited them in May 1969. The machinery for canning stood idle and a little out of date; the only activity was the salting of sardines, a delicious Dalmatian speciality. After salting, the fish are sold in small wooden casks, and some are now exported; 500 barrels were sent to France in 1969 as a first order. The two workers, one a woman who sang in the hard, throaty tones of centuries-old folk-lore as she cleaned fish, seemed burnt up from working with salt. The man showed us round, speaking in half-words of dialect in which I had to search for a few key words of understanding. To salt the fish, coarse sea salt is mixed with seawater and the fish are packed in it as tightly as possible in the barrels. When the fish are cured the flesh is quite dark and soft; it can best be eaten with head and tail held in the fingers, when the flesh comes straight off the bone.

It was a warm sunny day with a breeze to ruffle the sea and turn it dark blue. The small *Kafana* near the quay had prepared *marenda*, a meal which replaces breakfast. Yugoslavs, particularly in the south, start work early, sometimes before six in the morning.

They rise to Turkish coffee and perhaps *Šlivovica* (a plum brandy) or, more often in the islands, *rakija*. Normally, they eat nothing until 10.30 when they tuck in to a fair-sized meal. On Susak this consisted of new bread, cheese, salami, gherkins, black coffee, salted sardines and *Šlivovica*. *Marenda* sees the Yugoslavs through to the end of their normal working day, about 3 o'clock in the afternoon, when they have their largest meal. An interesting peculiarity is that they drink cognac as an aperitif and not after the meal.

Fortified, we staggered up the steep street towards the church and fortress at the top. The road finally turns itself into a flight of broad steps with gardens bordered by bamboo poles on either side; beautiful and quite large lizards scurried about. Bamboos grow freely in the sand of Susak and are used for stockading gardens, guarding vineyards, poking the fire and anything else that comes to mind!

The little church of Sv. Nikola is simple, airy and cheerful. It was full of light on that Saturday morning, and full of flowers as well. The bishop had paid a visit on the previous day, and four or five children were to receive their first communion on the following Thursday. The bishop's red silk chair stood on a rich Bosnian carpet. The altars were laden with flowers (white marguerites and red pelargoniums), while maidenhair fern climbed round the roof in front of the main altar. Under the statue of the Virgin was the greeting *Zdravo Marija*. *Zdravo* is used as an ordinary greeting when Yugoslavs meet, almost like our 'hallo', and seeing it in church makes religion seem a very friendly affair. The church is very old and stands beside the remains of an eleventh-century Benedictine abbey; a cross of the same date is built into the wall above the side door. Inside there is an enormous crucifix, dating probably from the twelfth century. The cross itself is made of black wooden planks nearly as high as the church. The black-bearded figure of Christ is a primitive wood-carving; his face is that of a sailor; the strong muscles are well carved and the feet have long toes. The whole is painted in strong, living tones. This crucifix floated in from the sea some fifty or sixty years ago, after a violent storm, and it is believed it was washed overboard while in transit by ship. Since its arrival on Susak, the Italians have made several representations to have it removed to Italy as its place of origin. These have been fiercely resisted by the islanders. They claim that although even now the crucifix is extremely heavy and almost impossible to move, in its waterlogged state it was heavier still; yet the men of Susak were able to carry it up the steep hill and place it in their church. This, they maintain, proves it to have reached their hilltop church by some almost miraculous strength — sufficient evidence that it belongs there.

After looking at the remains of the fortress, which commands a splendid view of the sea, we entered a small whitewashed cottage. In a corner of the living-room was a wood fire, under a hood and chimney, heating the stone sides and floor of the bread oven. The couple, in their sixties, were going to America to visit their children, and mama was preparing dough for the home-made Susak bread they intended to take with them on their journey; no modern, steam-baked bread for her! She kneaded the dough in a great wooden trough like a primitive cradle and added a little good olive oil to the flour. As she worked, her dark blue eyes darted about; she was as bright as a plump little partridge. Her husband, who had always been a sailor, had the thicker features of the seaman, with eyes half closed from salt spray. They were leaving for one, two or three years, they said. 'And will you come back?' 'Of course. . . .' I remarked that they must always come back because Susak would pull them home. 'Ah, *gospođa, vi dobro poznate Jugoslovenske ljude* . . .!'[1] And, indeed, it is true; though emigration means money and an easier life, Yugoslavs, and particularly Dalmatians, are seldom completely happy when divorced from their own soil.

It might be thought, because of the island's isolation, that disease would be more prevalent on Susak than elsewhere, and certain unhealthy characteristics exaggerated. The survival of primitive customs, intermarriage, etc., make the island ideal ground for a detailed analysis of health, and a careful survey was carried out by the Yugoslav medical authorities over a number of years. Contrary to expectation, it was found that the islanders enjoy as good, or better, health as the people of other

[1] 'You really understand Yugoslav people!'

similar but less isolated regions. The report summed up: 'Men and women have merry temperaments. Fond of songs; simple, realistic, and of a conservative turn of mind; are much attached to their customs and adhere to ancient moral principles. Are distrustful at first, but later very kind and hospitable. Fond of their vineyards for these are their life essence.'

Among the conclusions reached by examination and analysis were the following. The death rate is slightly lower than in other areas; hygienic conditions about the same. Nutrition was satisfactory as regards calories but deficient in protein, in spite of much meat and fish; no alcoholism, though the amount of alcohol consumed by males is much greater than the national average. Intestinal diseases are no more frequent than in other regions, despite primitive lavatory conditions. No trachoma. Tonsil hypertrophy frequent, but only one case of grave multiple malformation. Venereal infections extremely infrequent except where members of a family have spent many years abroad; syphilis is called on the island 'American Disease'.[1] There is, however, a high proportion of damaged teeth, probably due to a lack of fluor in cistern water. The report was prepared in the 1950s.

Susak, as can be seen, is the plus-individual island in what is a highly individual part of the world. Tourism, as yet, has hardly touched it but, whatever its future, the strong personality of the island of Susak can be relied on to guard its unique character.

[1] Author's note: It should be mentioned that 'America' usually means any country to which the Dalmatians emigrate for work and could equally refer to Australia, New Zealand or Canada.

Chapter 3 The islands of Rab and Pag. Zavratnica Fjord

I am often asked which, of all the Adriatic islands, is my favourite. It is an impossible question to answer when the choice is so vast, the differences so great, and the impact of personal experience so varied. I can only say that perhaps the island to touch me most deeply with its beauty, its helplessness and its transparent gentleness, is Rab — lovely, sad-eyed Rab. But make no mistake, Rab is not sad-eyed with self-pity. On the contrary, she offers gaiety and hospitality and the freedom of her splendid island with open-handed generosity to all who arrive on her shores. The sadness is that of a Mona Lisa, and lurks indelibly under the smile, painted on by the cruel brush of history.

The island of Rab is less than 100 square kilometres in area and lies north-west to south-east, with its most easterly tip close to the mainland. Its north-westerly end is well out to sea between the bigger islands of Krk and Cres. Like them, it is one of the Kvarner group. It is served daily by passenger ship from Rijeka, and there are also car-ferries from Jablanac to the town of Rab and from Senj to the village of Lopar. Every year 100,000 tourists land on the island, and in summer the little medieval town of Rab buzzes with gaiety, its streets crowded with bubbling tourism, and the main quay, when the ship comes in, a milling conglomeration of tourists meeting friends, and hotel receptionists searching for their new charges. Away from the town, it is easy to find solitary corners and coves, and little-frequented villages.

The island has shores the colour of strong jewels — deep sapphire and emerald — chrysoprase pinewoods, white and black rocks, like onyx under a backcloth of deep blue sky. Pliny called Rab Oppidum Arba (it was known to the Romans as Arba) and, indeed, before the coming of the Venetian overlords, with their expensive appetite in wood and monopolies, Rab must have been richer in beautiful trees than the neighbouring islands were. Her greenness is preserved by the Kamenjak mountain stretching along her eastern side, which protects the rest of the island from the north-easterly *Bura*. Kamenjak (408 metres) is known affectionately by Rab people as their 'Protector'. The island never lacks for water; there are many surface springs and over 300 wells. Its climate, too, like its nature, is benign. The average temperature in January is 6·7° C., and in July, 23·2° C. One of the sunniest spots in Europe, it has an annual average of 2,500 hours of sunshine.

Originally, Rab was inhabited by Illyrian Liburnians. Later, the Greeks formed three settlements, one of which has left massive remains on the heights overlooking Barbat. For those who like a stiff climbing walk, an expedition to the top of the hill to Sv. Damjan, where this Greek town was located, is well worth while, especially if, after descending to sea-level again, they order a meal at the fish restaurant and cool off in the sea while it is being cooked. Beside the small restaurant's jetty are boxes full of live crayfish, and ropes heavy with farmed mussels, which cling obligingly to the sisal strands until a client orders *moules marinières* or *risotto mužule*.

Rab was first recorded in 360 B.C., and the Romans arrived during the second century B.C. The city walls of Rab town, massive portions of which can still be seen and even walked on, were built about 150 years *prije naše ere*, before our era, as the Yugoslavs call 'B.C.' The Romans — and who can blame them in such a lovely place? — built various country villas on the grand scale. Later, Rab became part of the Eastern Empire; in the ninth century part of the Croatian kingdom; and, with a few minor vicissitudes, in 1409 Venice overran and ruled it until the Austro-Hungarian Empire took control in 1797. Napoleon annexed it as part of his Illyrian Provinces in 1805, but in the debacle of 1814 it was returned to Austria who remained in possession for 104 years. Even in 1918 Rab's subjugation was not at an end; Italy swept in until, under the Treaty of Rapallo, Rab became part of the newly formed kingdom of the South Slavs, 'Jugo-Slavija', *jugo* meaning 'south'.

Rab town has a population, apart from tourists, of about 1,500. The old part, one of the most attractive anywhere, is built on a sharply rising peninsula-promontory between the large harbour on its eastern side and Sveta Fumija Bay (St Euphemia, a historical heroine) on the other. It is a very small town, even by medieval standards, and has three main streets running parallel one above the other on the steep incline. Traversing these main streets, which are paved and devoid of traffic, steep and narrow passages run uphill to the top of the promontory. The town itself is like a small jewel composed of many stones, all perfect in their way. Palaces, churches, museums, Gothic and Renaissance façades, ruined towers and crumbling Roman walls, and magnificent carved stone windows and balconies — all reflect a golden sunlight. It is ancient, ageless, yet teeming with life. Butchers' meat can be dimly seen through Venetian palace doors; modern household appliances gleam through ancient arched windows; hand-embroidered souvenirs are held for sale in the gnarled hands of old grannies, while young Rab girls flirt their short skirts as they disappear through Venetian archways. There are so many outstanding medieval buildings in the town that to list them would be tedious, and description inadequate: far better to visit, and marvel at their mellowness.

During the reign of Augustus (27 B.C. – A.D. 14) Rab was a fortified base for the Roman fleet. In the early Middle Ages it became the seat of a bishop. The first mention of the bishopric was in 530, and the bishop's name was Ticianus. His signature appears on a document signed at a meeting of a council at Solin, near Split. The see of Rab lasted until the early nineteenth century, and the last bishop, Galzinja, was enthroned in 1820. In 1828 the episcopal centre was moved from Rab to the island of Krk, where it still remains.

Among all the buildings, the best for me is the little cathedral of Sv. Marija, consecrated by Pope Alexander III in the twelfth century. Now reduced to the status of a church, its outer walls overlook the vertical cliff above the sea of Sveta Fumija Bay. The first time I went there was in 1963, and I shall never forget that first impact. It was a Sunday morning and the small building was packed. There were old men and old women in peasant-black; young women in brighter colours with large-eyed children, some clinging to skirts, and some too young to walk. The faces of the older women were permanently creased in the searching, faraway look of those who continually gaze seaward for a glimpse of a returning fishing boat: patient, anxious, but un-despairing. Here, it seemed to me, lay the true essence of Rab.

The present building was first mentioned in 1118, but there had been a church on the site much earlier. The beautiful and ornately carved stalls, gleaming darkly against the stone fabric, were made about 1445. Outside, the belfry stands alone like an uncompromising square finger, stretching up into the sky. It dates from the twelfth and thirteenth centuries, but in 1495 the top collapsed and was rebuilt by Juraj Branković, a master mason of Šibenik. While Rab remained an episcopal see the church was rich, but later it became poor, losing both lands and revenue.

Outside the church door a shady tree flourishes. It is a holm oak planted in 1921 to celebrate Rab's freedom under the Treaty of Rapallo.

There are several convents, some of enclosed orders; but in one, that of Sv. Antun, the nuns showed me a large oil-painting 400–500 years old. It depicts the Holy Family up in the clouds above the medieval town of Rab, which the artist painted in considerable detail. And it is a fascinating record of what the old town looked like. Rab has always attracted painters, and two modern artists, both born in Rab, have their work on permanent exhibition there: Pik Bolković and young Matsolić.

Among the exhibits of a small museum on the ground floor of the priest's house are some icons, a Virgin and Child painted in 1350, a fifteenth-century tabernacle, a twelfth-century Limoges reliquary, a Roman 'family' altar in stone, a picture of Jesus and Mary, dated 1526 — painted by Juan Buscheta, a Spaniard whose tomb is in Rab — and a Crucifixion by Paolo Veneziano.

But the most precious relics are kept separately on the upper floors, in the priest's own quarters. He has a small and lovely garden, full of sweet-scented shrubs and climbers. He, like nearly all the priests I have met in Yugoslavia, is a kindly, human and humble man. Halfway up the open-air stone staircase he paused to pick some of the heady jasmine flowers rampaging against the garden walls. I waited on the shaded upper terrace while he fetched the treasures. The most important is the relic of St. Christopher, patron saint of Rab. The skull inside a double box is the colour of dark ivory and wears two diadems. One of these circlets is simple and frail with the fleurs-de-lis as its only decoration; the emblem serves as a reminder that in the thirteenth, fourteenth and fifteenth centuries Dalmatia was involved in the affairs of the House of Anjou, kings of Naples and Sicily. The second diadem was a gift of a queen of Croatia and is richly jewelled. A small jewelled cross contains, so they say, a fragment of the True Cross. But, as the priest honestly remarked, after such a long time who is there to prove that relics are the true originals or not; and, he added, what can it matter so long as people derive solace and comfort? Also in the priest's keeping are copies of the church documents which were kept in the cathedral until it was demoted in 1828. After the removal of the episcopate the original documents disappeared from Rab, one of them turning up in Leningrad! The island was an important centre both for religion and culture, and was an autonomous commune with a statute early in the Middle Ages. During the Crusades a number of Benedictine monks settled there. Earlier, King Peter Krešimir of Croatia, a contemporary of our William the Conqueror and one of Croatia's ancient heroes, donated two parcels of land to the church for monasteries, in order to 'promote learning and thought'.

In medieval times Rab suffered, in common with the whole of Europe, from plague. The worst epidemic was the bubonic outbreak of 1456. In those days the towns took stern measures, which included cleansing by fire. Towards the seaward end of one of the main streets there is an arch. At the time of the plague a solid wall was built at this point, closing the town from the outside world in an effort to keep the disease away. All those infected lived and died outside this point. The chapel of Sv. Nikola near the beautiful loggia was built in commemoration of the event.

During the Venetian overlordship Rab suffered severely. Her early economy had been substantial and sound, one of her flourishing trades being the production of silk from the many mulberry trees, and legend says that the original silkworm was brought home in a hollow bamboo by a Rab-born priest. A second large industry was the production of olive oil. But the Venetians demanded an annual tribute of wood, and without trees neither silk nor oil could flourish. Another hardship resulted from the curtailing of the salt-fish industry. The Venetian salt monopoly made it impossible for Rab economically to continue it and the fishermen fell on very hard times. Today the production of early vegetables, 1,000,000 kilos a year for the Rijeka market, together with shoemaking and a woodworking industry, help to balance her economy.

Apart from the town of Rab there are six other settlements on the island: Lopar, Mandanija, Supetarska Draga, Kampor, Banjol and Barbat. There is also, close to Lopar, a new tourist settlement in a beautiful corner which they are calling San Marino. The reason for this is that Lopar, situated in an attractive oblong bay, is believed to be the birthplace of the hermit stonemason who founded the republic of San Marino in Italy. Rab's San Marino is a kilometre or two away from Lopar across a narrow neck of the island, and faces the mainland. It has one of the biggest and broadest fine sand beaches in Dalmatia, under tall, pine-covered sandy cliffs. The beach is safe for children, and it is possible to wade out at least fifty metres. There is a huge hotel block with 1,000 beds, and the restaurant and reception areas are strikingly modern. These twin San Marinos — one on Rab, the second on the other side of the Adriatic, close to Rimini — are cultivating a reciprocal rapport, and organize excursions between the small republic and Lopar's new San Marino resort.

Two other places on the island of Rab draw me with a pull I find difficult to resist. They are close to each other; one is very old and the other quite modern. The first is the Franciscan monastery of St Euphemia, built at the south-eastern end of one of the most fertile and productive valleys in the island. The second, Kampor, is at the other end of the same valley.

The monastery was founded in 1458 and built in the grounds of an earlier foundation, the land being part of King Petar Krešimir's gift. In the mid-fifteenth century the Turkish invasion reached

Bosnia. Euphemia was the wife of a Bosnian knight and, when the Turks attacked her tribe, she travelled to the foot of the Velebit range of mountains — a very long journey for those times — bringing with her forty of her people. She then waited three years for permission to come to the island of Rab. Finally, the Duke of Rab allowed her to bring her Bosnian followers and make a new home, admitting on the island only those whose names ended in 'ić'. As a thanksgiving Euphemia built the monastery.

The monastery is a quiet and gentle place. The Franciscan Brothers go about their daily life in the delightful matter-of-fact way which one finds in Yugoslav monasteries. A fishing-boat floats near at hand ready for use, and the garden is full of sensible crops. In the church — Gothic, and reconstructed in the sixteenth century — there are fragments of interlaced stone carving from the ninth and tenth centuries in the choir screen; in the side-chapels there is a polyptych done by the Vivarini brothers in 1458, and a late Gothic wooden cross of great beauty. In a big room at the side of the cloisters a small museum contains a splendid mixture of Roman, early Croat, and late medieval bits and pieces; carved stone, coins, illuminated manuscripts and life-sized domestic tableaux jostle for the visitor's attention. There is a benignity in the atmosphere at Sv. Euphemia for humans, birds and insects alike. On the steps outside I saw the biggest and most handsome green and yellow moth I have ever set eyes on, like a lepidopteral tank in camouflage.

The second place, Kampor, has a stream running through its centre and is again, since the war, highly cultivated. During the war it was made into an internment camp for Slovenian families whose sons had been accused of guerrilla activities. On the fields and vineyards massive concrete living-quarters were built, some of which can still be seen. At Kampor there is now a cemetery, without individual graves, but with little plaques bearing the names of those who died in the camp. The names came from the church register because the Italians were obliged to record the deaths, of which there were more than 4,000 out of the 8,000 or so who were imprisoned.

The impact of this cemetery is quite overwhelming — a monument to the inexplicable atrocities of war, and of man's inhumanity to man. The dead are those who did not survive the rigours of the camp. At first the interned were treated fairly. Then some quislings were introduced by the Italians. When the interned men and women discovered the traitors they killed them. From then on the treatment changed. Civilian mothers and fathers, wives and small children — the Italians did not bother to kill them. Instead, they let them starve and gave them no water, though every man, woman and child could see the stream of fresh water, which was barred to them, running through the centre of the camp. The dead included fourteen babies who, in one terrible night of storm and rain, were drowned by flood, as if even nature were showing a malevolent humour.

The cemetery and memorial have the greatness of complete simplicity. At the entrance, plain walls built of chunks of stone enclose a small square. Three plain, thick stone columns stand upright with simple words in Croat and Slovenian:

'If your eyes are blind with tears
And grief darkens your face,
Then, Mother, don't look for my grave.
This free land is a living form of your son.

Something easier for you:
This small clod of earth
Feeds and protects you as your mother would;
And, even in the darkest hours, embraces you
Like a living son, giving you life.'

At the farther end of the courtyard, an opening leads to the rows and rows of neat white stones and name-plates. It is all kept as a green garden with shrubs and trees and grasses. In the centre there is just one gigantic quadrilateral column of stone blocks, rising into the sky like granite links in a solid chain. A broad dividing walk runs from end to end of the garden, and even this is symbolic. Of white marble, it is striated with diagonal lines on each side of a broad central channel to represent the flow of running water: a reminder of the prison camp's stream which could never be reached.

Many Italians come to Rab as tourists; when they hear of the cemetery they are shocked and horrified, and hate to talk of it. What is it that changes men so utterly in time of war?

In the villages of Rab, life continues much as it has through the ages. There are donkeys, and women who tend their flocks while spinning yarn on long spindles from sheep and goat wool. Wine is made from family vineyards, and some say that the best Prošek, always an outstanding dessert wine, is found in Rab cottages. After the tourist season, the entire island sinks back into quietness and unchanging custom: a perfect place for work, lotus-eating or solace.

Zavratnica Fjord

For anyone taking a car to Rab on the ferry from Jablanac, an hour or two to spare gives the opportunity to see the remarkable Zavratnica Fjord. Just south of Jablanac, this small seawater inlet, deep and quiet, has been made into a Lilliputian national park. From the car-ferry quay at Jablanac, a narrow and rocky path skirts round the fjord. Above the path, the steep limestone karst reflects a white warmth. Guarded by its status, like our National Trust areas, it is forbidden to fish, or pick the flowers which cling to crevices close to the path. The water is so clear, so deep, and so full of fish of all varieties that it is like visiting an aquarium, only here it is the fish which are unrestricted. Big fish and little fish go about their business undisturbed by human beings, or even their shadows. A flash of silver shivers from the depths as a shoal of small fish turn, with acrobatic precision, before the approach of bigger ones; the superior nose of some master-fish will drive a small squid nervously into his crevice. Marine life in its blue and silent world lies, natural and unperturbed, spread out in the depths before one's eyes; a revelation of the second world which is always close at hand in the Adriatic.

Pag

The island of Pag, long, thin and close to the mainland, sprawls from Rab, which it overlaps, in the north, to Zadar in the south. From a distance it looks bleached and skeletal, sucked dry by the gods and abandoned. No one, even in their wildest moments, could call Pag a lush island. The section nearest the mainland of this long sliver of limestone is a sun-cracked and barren waste of creamy karst, with hardly a scrap of green. Eroded rocks stick out from the slag which has fallen to the edge of the deep, almost Prussian blue of the sea. Not a sign of life, beyond a few gulls, relieves the primeval stillness of this inshore ridge. Yet the other side of this barrier, on parallel ridges, there is enough vegetation to support 30,000 sheep, sundry donkeys, a few ponies, a large number of keeper-reared pheasants, partridges, and hares, and a population of 9,000 in the three main villages.

Pag is not very big, covering 284 square kilometres — just over half as large as the biggest island, Krk. Yet, because it is so long and thin, its vegetation varies considerably between one end and the other, and one almost expects climatic variations between north and south. Certainly, the inhabitants vary from one end to the other: history has had a hand in this.

Although it is true for much of the Dalmatian coast to say that the Venetians were responsible for denuding the slopes of wood, there are sometimes other reasons for the bareness of the rocks. Pag is a classic example of the second reason — a climatic one. In winter, and sometimes in summer, the powerful *Bura* smacks down from the Velebit mountains on to the sea, picks up the salt spray, and deposits it on the exposed ridges. No plants, or even trees, can thrive with this treatment, and any loose top soil is soon blown away. All that is left is the creamy-coloured stone on which not even goats can graze. So the approach to Pag, whether you make it by ferry from Karlobag or drive over the awe-inspiring new bridge from Zadar, is startling; rather like the approach to a lunar world.

Historically, Pag has been rich though it now seems half forgotten. One reason for its early population was undoubtedly the presence of fresh-water springs at Pag town, Novalja, and Metajna.

In prehistoric times the island was inhabited by Illyrian Liburnians; there are still the remains of fortifications and burial grounds near Novalja, on the ridge between Novalja and Pag town at Kolan, at Dabovo, and at Vidasov. Although no traces of early Greek occupation have yet been found, people on Pag are inclined to the theory that 'Pagus' was a word given by the Greeks, meaning 'water-source' or 'spring'. The Romans, early in the first century A.D., if not before, set up a series of fortifications on the island as a protection against the Illyrian tribes whom they had pushed back, so that the Roman province of 'Illyricum', as they called these Dalmatian lands, had to be distinguished from the area the Illyrians still held, and the division became Illyria Romana and Illyria Graecia. There were several Roman forts on Pag island — at Košljun and Zaglav (near present-day Pag), the large Castrum Cissa now called Časka, and the naval fort, Castrum Novalia, the present-day Novalja. Besides the strong-points there were sizeable settlements at Pagus, now known as stari Pag, at Tovarnele, on the seaward side near Lun, and other places near Novalja.

The Slavs moved in early and made their main base at Kissa, Kesa, or Novalja, as it is now called. During the Middle Ages, the long, narrow island was carved up to suit the appetites of Dalmatian events. In 1071 King Petar Krešimir divided it in two, giving the northern half to the Church of Rab, and the southern part to Zadar. This led to constant clashes between Zadar and Rab. One prince of Rab signed himself '*Ja Mazzolin, Knez*[1] *Rapski* (of Rab) *i Časki* (and of Časka)'. An astonishing state of affairs in the days when insular defence and unity really meant something! No wonder the nature of the divided people grew apart — a fact which is still apparent today. When Kissa was destroyed in the twelfth century, stari Pag became the civic centre. Between 1111 and 1154 Rab and Pag and the town of Zadar came under Venetian rule as the result of numerous battles and clashes between Venice and the Hungaro-Croatian kings which had been going on since the days of Krešimir. During those early years the islands, as they were again in recent times, were pawns in the bigger power-games of Europe.

So Pag became Venetian for a few years. In 1174 Count Dominik Mauroceno, who had been very helpful to Venice, was made a present of half the island of Pag: 'everything from Kissa to the fields of Lun'. This did not suit Rab at all because, though she too

[1] Prince.

was dominated by Venice, Lun was still part of her diocese. In 1192 Duke Michael quietly annexed the town of Pag from Zadar, too, and the island was in one piece again. In the middle of the next century King Bela IV made Pag a free, royal town on 30th March 1244, in recognition of the help Pag people had given him during the Tartar invasion. Pag became rich with her salt-pans, her olives, and her famous cheeses and, following an uprising against Zadar, again achieved partial autonomy in 1311. Finally in 1376 King Ludovik I[1] gave the town full autonomy and equal status with other Dalmatian towns, and the island became one unit, but only for six short years, from 1376 to 1382. Then once again it was divided between Zadar and Rab. In 1409 Pag came finally under Venice and her municipal statute was confirmed in 1433.

All that remains of the original thirteenth-century town of Pag is the church, the walls of a few houses and the remains of a sixteenth-century Franciscan monastery. The church is simple with some lovely stonework and carving. It is thirteenth-century Romanesque-Gothic. Sadly, the wooden lining of the ceiling is coming down in places, and the stone tracery of the rose window is being eaten away by salt picked up in the wind from the salt-pans near by. The church has a forlorn and forsaken air about it. A huge wooden altar-screen rises to the roof; and the famous statue of the Virgin and the infant Jesus normally rests inside this church. This is a poignant thirteenth-century composition richly coloured in blue, gold and terracotta. Twice a year the forsaken church comes into its own: on 15th August, when the statue of the Madonna and Child is carried in procession to the present-day town and cathedral half a kilometre away on the other side of the salt-pans, and on 8th September when it is taken back again until the following year. The legend is that after Pag suffered a particularly severe epidemic of plague during the Middle Ages, the town elders vowed to process with the statue every year. I hope that, before long, someone will vow to restore the church the statue rests in. Perhaps the coming of tourism will provide both the will and the means.

[1] King of Hungary and Croatia (sometimes called 'Louis the Great') was of the House of Anjou.

The new town of Pag was begun in 1443 after the old one was burnt to get rid of the plague, and built to the north of the salt-pans. It was designed, probably, by Juraj Dalmatinac (George of Dalmatia) — that inspired architect and stone mason — with a broad longitudinal street and regularly placed transverse thoroughfares. The cathedral, begun at the same time, took eighty years to complete. In the courtyard of the duke's palace there is a well with carved stone sides about a metre high. If you look inside, you can see the grooves worn in the stone by the ropes pulling up buckets. The cathedral, the Palace of the Podesta, and the Bishop's Residence, all owing their inspiration to George of Dalmatia, are in the main square which is best approached from the harbour by George of Dalmatia Street. Facing the street is the cathedral with its beautiful carved figures, its rose window, and its arched double doorway. Angels in the arch over the doors are deeply carved in bas-relief, and in and about their wings swallows build their nests. Juraj Dalmatinac, who carved so much of life into his stone, would, I am sure, have been happy to see them living with his work.

Today, the walls which used to surround the town have almost disappeared except for a tower near the harbour. Originally, the only house outside the walls was the harbour-master's, which served as a quarantine building; all the townspeople and the domestic animals lived inside the walls. Cholera was frequent, however, and people began to think that the narrow streets and the presence of the animals might be the cause, so they pulled down the town walls and used the stone to build new cow-houses higher up the hill.

The approach to present-day Pag gives no idea of the beautiful city which lies behind the harbour. The quayside buildings are nondescript and the townspeople take for granted both the old and the new. Pag citizens have forceful personalities and strong features. The women have long been famous for their very beautiful and delicate pillow lace; it is still a thriving cottage industry. Pag lace edges their white linen coifs and makes a fine foil for the coloured satin aprons and braided velvet jackets. The other product which has been renowned since Pliny wrote about it is Pag cheese, made from sheep's milk. Powerful and extremely good *Paški sir* finds its way to many parts of Yugoslavia. The salt-pans yield about 12,000 tons of salt a year — sometimes more if the summer is very dry and hot — and form Pag's main industry.

The road to Novalja skirts the other side of the shallow lagoon on which Pag lies and, once past the modern tourist hotel, continues as a narrow dirt road. The earth on either side is poor and sandy but does grow good wine grapes. Small stone huts, scattered here and there as shelters, have rush roofs which are tied down with wire. Some of the fields, enclosed with stone walls, seem too bare even to support one goat, yet sheep are grazed on the few prickly plants and produce the famous cheese. About seven kilometres from Pag there is a wayside shrine, and behind it lie the remains of a very early church, half hidden by the windswept scrub. The trees on this bare ridge are bent almost double by the wind, but suddenly the road plunges into a sizeable wood of scrub oak, *primorski Hrast*, which, by comparison, seems lush and almost sub-tropical.

The small port of Novalja has one very attractive-looking modern hotel, built round a formal garden with a fountain at its centre. But the port itself has nothing modern about it, going gently about its own business. I was there in October, and the scene reminded me of a clean-washed watercolour. Boats were unloading mounds of black grapes, while a large and very black tame jackdaw was holding the centre of the quay, strutting on a piece of ornamental ironwork. Near some cottages an ancient and communal wine-press stood ready for use by Novalja families.

Modern Novalja has been built on the Roman town which served as a port for the garrison at Castrum Cissa. A first-century Roman underground water system is still in a good state of preservation. A little to the north-east there are some walls and a mosaic floor of a fifth-century church. Also near by is an ancient stone quarry with indications that work was suddenly interrupted in the seventh century, perhaps when the Slavs first appeared.

At Stara Novalja, five kilometres north, there are still, under the sea, the remains of ancient walls belonging to a yet earlier settlement; and on the edge of the plain more walls and the remains of a necropolis.

Beyond Novalja, still travelling towards Rab, the rough road becomes worse, the stony fields sparser until suddenly the land changes and olive trees make their appearance. Each tree on this crest has a protective circular wall of dry stones serving the double purpose of protecting the young growth and clearing the immense proportion of stone from the available earth.

The port of Lun lies at the foot of the steep hill. The community lives almost divorced from the rest of the island by both bad communication roads, and also by its own inclination — built up through the ages — towards its spiritual cousin, Rab. The people of Lun are close to Rab, both physically — 4½ nautical miles, as opposed to 11 from Novalja — and in mentality. All the Lun boats are Rab-registered. All their wine and other produce goes to Rab to be sold. Even the lighthouse belongs to the authority which looks after Rab lighthouses. The population numbers little more than 300. For holiday-makers who like to savour local life in a forgotten and beautiful corner, Lun is an experience to remember. There are 70 private rooms for tourists and a small cottage restaurant under some trees. The wine is good, the fish is good, the pršut is extra good, and life is easy.

Chapter 4 The North Dalmatian Archipelago The islands of Ugljan, Dugi Otok, Pašman, Iž, Zlarin, Murter, Prvić, Žirje, Primošten, Visovac, etc. The mainland towns of Zadar and Šibenik

South-east of Pag, indeed almost due south, are a group of islands offshore from Zadar. Although not particularly small — Dugi Otok is the largest (124 square kilometres) — they are of a very different quality from Rab, Pag and Lošinj. The villagers look to Zadar for a lead and are not very rich. Zadar has always been irrepressible; turbulent, determined, awkward at times, and stubbornly set on paddling her own canoe whenever possible. Maybe it is this Zadar strength of character, reminiscent of a forceful brother, which has made her attendant islands take a secondary role, like satellite sisters to the main star.

The nearest island is Ugljan, facing Zadar, and about twenty minutes away by motor-boat. To the south-east is Pašman, a little bigger and separated from Ugljan by a very narrow channel. The other islands of this Zadar group of the North Dalmatian archipelago lie in thin parallel lines further out to sea. The middle row includes Olib, Molat, Sestrunj, and Iž, while the outer line has Premuda, Ist, Dugi Otok, etc., and extends as far as the Kornats. Altogether the Zadar group includes over 150 islands, if you count the rocks and reefs; only 18 of them are inhabited.

Small motor-boats, privately owned, as well as small Jadrolinija passenger/cargo ships, make the journey from Zadar to Preko harbour on Ugljan frequently during the day. If you just miss a regular ship, then you get a lift on a small open boat. The first thing you notice on the approach to Preko is the large medieval fortress of Sv. Mihovil at the summit of the steep hill, 265 metres above the harbour. It was originally a monastery built by Benedictine monks, possibly as early as the tenth century. In the thirteenth century the Venetians made it into a stronghold and lookout point; it is a splendid spot for a panoramic view of all the surrounding islands. The second feature catching the eye is a tiny round island, thickly covered with trees, grass and flowers, just outside Preko harbour. The water between is shallow and calm, almost like a lake, and a man can swim across without going out of his depth; in fact this is the traditional place for Preko children to learn to swim. The island is called Galovac, or sometimes Školjić, the former name probably coming from 'Galle', the name of a Zadar family who owned it in the fourteenth century. A Franciscan monastery and its surrounding gardens completely cover Galovac. Sylvan paths make a complete circuit close to the sea. The monastery was built in 1443 on the site of an earlier hermitage and was reconstructed in the sixteenth century and the monastery church was built in 1569. Before the war many Preko people made a habit of going to church on the island by boat, and it was a Sunday family outing. Now there are only two Brothers in the monastery, but they keep the gardens beautifully in the Adriatic manner. There are pine trees and cypresses, with patches of iris and other bulbs, and flowering shrubs between stone borders, with parts near the sea shore in a semi-wild state. The Brothers gladly allow tourists to spend the days sunbathing and swimming; it is quiet, peaceful and beautiful. The monastery at Galovac had a printing-press which was used during the first patriotic stirrings of independent thinking in arts and politics during the Austrian oppression of Croatian culture. In the mid-nineteenth century Austria had discouraged progressive thought, suppressed the freedom of the press, banned progressive newspapers, closed the public reading-rooms, and prohibited use of the national tricolour flag. Both administration and schools had to use the German language. In the small graveyard, among the tombs of former Brothers, one of a group of men who worked towards a greater freedom for Croatia is buried. Somehow, a little of the sadness of those times seems still to be attached to this grave.

Preko (the word simply means 'across' from Zadar), in spite of being the largest village on Ugljan with a population of over

2,000, is quiet and restful, and so far tourism has touched it only lightly. All the inhabitants know each other well, and half of them seem to be related. One of the biggest families is Sorić, and Sunday lunch with them can mean sitting with 20 or 30 brothers and sisters, cousins and grandchildren; their immediate family number 58. Preko was probably a village during the Roman era and the whole island has been inhabited continuously since the late Neolithic Age; there are numerous Illyrian earthworks as well as Roman remains. The older men and women in Preko and the other villages prefer to wear traditional peasant costume, though their children may lead a sophisticated life in Zagreb or Belgrade. Many of the sons and daughters go abroad, and occasionally the parents visit them. When they do, they change their Preko dress for the smart trappings of the West, but on their return they go back to their black clothes and kerchiefs. But they are pleased when the younger generation is successful in the 'bigger world', and the whole community celebrates the fact that Ante Sorić, one of Preko's 'boys' has recently been elected as member of the Croatian parliament representing Zadar and the local islands.

Offshore and a little to the south-east of Preko is another small island, Ošljak. During Venetian times it was used as a leper colony or lazarette. It is an attractive place, with some old houses perched on its steep face overlooking the water. The Zadar boat calls there on its rounds. Ošljak faces Kali. This small port of Ugljan is famous for its fishermen, who can be found all over the world. They have a tradition in fishing; it is their whole life, and they are not interested in tourism. Still further south-east is the port of Kukljica, in a sheltered bay. Unlike Kali, Kukljica has opened its doors to tourism and, on one pine-wooded arm of its wide harbour bay, has built an hotel and some new holiday bungalows which have the amenities of hotel service. But Kukljica is still very much a fishing port. It is also very close to the south-eastern end of Ugljan and the narrow sea passage which separates it from Pašman. There is little pasture or woodland at this end of Ugljan, and the main crops are vegetables, olives and figs.

The narrow water separating Ugljan and Pašman is 20 or 30 metres wide and has only been navigable, even for small craft, since an artificial channel was dug in 1883 and called Zdrelac. There are posts in the water to mark the channel, red to port and black to starboard.

Pašman (57 square kilometres) is, geologically, in two parts. The western end consists of chalk limestone, while the eastern part is of Dolomitic rock with a small area of fertile diluvial soil. While the limestone area is mostly under maquis growth, the Dolomitic and diluvial zones are well cultivated into vineyards and vegetable plantations. The steep western coast with its limestone rocks has a whole collection of coves and indentations which are completely protected from the *Bura* wind. They make a wonderful playground for holiday-makers with small boats and a taste for fishing. The most westerly village on Pašman is Zdrelac. In the first week of May, when I was there, the field flowers were out and the insects busy. The broad glades and fields behind the village were dotted with sheep grazing under fruit trees; it seemed one of those perfect and eternal settings for Shakespeare's *A Midsummer Night's Dream.* Round a point is a curved sandy beach, sheltered and warm with smooth fine sand — an amenity not always easy to find where even gravel is termed 'sand' in translation.

There are no roads on Pašman, only small stony lanes, often between dry walls. A small boat is the easiest way to visit the villages, eight in all, the most southerly of which is Tkon, opposite the mainland town of Biograd. The people of Tkon are fishermen, but they also produce wine, olives and figs. On Čokovac, a hill to the north-west of Tkon, there is a Romanesque abbey, a Benedictine foundation built by the monks when they fled from Biograd in 1125 as the Venetians sacked and captured the town. Biograd, at that time, was the important civic centre of Dalmatia; King Petar Krešimir of Croatia founded a bishopric there, and monasteries and convents in the middle of the eleventh century. Then King Koloman (Rex Hungariae, Croatiae et Dalmatiae) was crowned King of the Croats in 1102. On the one hand, Krešimir had connections with the Orseoli family in Venice; on the other, Koloman was of the royal Hungarian line. No wonder there was trouble with Venice after the Magyar take-over!

From Pašman one can either go out to sea towards the southern end of Dugi Otok or, as I did, return to Preko and approach the other islands of the group from the north-westerly end. This, for me, is the better way. The inner islands are so peaceful. The water is pale in the evening light and a rich dark blue in the morning, and one can spend days just moving from village to village by small boat. In the evenings schools of porpoises dance their water ballets within a hundred metres of the boat, while cormorants look sideways at their antics and continue fishing. The waterside villages welcome the stranger in the old manner of pre-tourist days.

The island ridges show big patches of cement on their slopes, put there as catchment areas for rainwater which is channelled into underground reservoirs for the village wells. The villagers are great gardeners and grow not only vegetables but masses of flowers round their doors, often in painted oil-drums. Tomatoes and lettuces are protected from too much sun or too much *Bura* by standing branches cut from the wild maquis scrub, and everything is frequently watered so that the growth rate is tremendous in the warm climate and vegetables are very tender, like hothouse produce. Apart from the cement catchment surfaces, the ridges are vertically striped with dry stone walls, thick and solid, which from the sea look extraordinarily regular. The men of these islands hold their land very dear, and each peasant family owns various patches on the hills which they mark out with the stones they clear from the fields.

At the northerly end of Ugljan is the village of the same name with a population of some 1,400 and a small harbour. An old Franciscan monastery dating from 1430 stands at the entrance to the harbour. In its courtyard is an epitaph to Bishop Šime Kožečić Benja who was the founder of the Glagolitic printing-press in Rijeka; he died in 1536. Near by are the remains of a Roman rustic villa and, further back in time, a prehistoric earthwork.

Beyond Ugljan another hamlet, Sušica, is completely on its own. I was there on the first of May, a national holiday in Yugoslavia, to meet some friends and have *marenda* with them (*marenda* here usually consists of wine and coffee, ham and cheese or salami taken at 10.30 in the morning as a substitute for breakfast). This particular family typifies the political predicament so many found themselves in between the wars. After the Treaty of Rapallo (12th November 1920) Zadar (like Istria and the island of Lastovo), went to Italy. Although the islands outside Zadar were Croatian, life for the islanders was difficult. With Zadar in foreign hands children either had to attend school in Zadar, learning everything in Italian, or obtain a Croatian secondary education further afield. Many young men emigrated and some turned to political causes, among them communism which, at that time, was punishable by imprisonment. The family with which I had marenda had produced sons in both categories. My host, now aged sixty, had emigrated to America while his brother had suffered in the 1930s for being a communist. There is now a plaque in a nearby wall commemorating him.

Not only the men emigrate, but often younger women leave as well. My host related, with many guffaws and in front of his wife, the story of a girl friend he had had in his youth. She had gone to the United States to work for several months. When she came back her speech was full of Americanisms such as 'O.K.', 'Yeah', 'So long, Honey', etc. He mouthed all these at us with much pantomime. She had also put on a great deal of weight and was enormous. He had lived with her for four years, intending perhaps to marry her, 'but', he said, 'it was all too difficult; she was too big, and so awkward to get on top of'. He howled with laughter, and so did his wife and daughter, standing behind him. Though women on the islands are still subservient to their men, this kind of peasant logic and practicability appeals to them all.

Beyond Ugljan are the islands of Olib, Silba and Sestrunj. From the sea Sestrunj appears uninhabited. This is because the only village was built in the interior, which was safer in the days of marauding pirates.

Silba (14½ square kilometres) also had only one village, of the same name, with a population of about 400. At the end of the eleventh century it belonged to a Benedictine foundation in Zadar. Later it was leased by the Venetians to individual Zadar families. It prospered during the seventeenth and eighteenth centuries as various rich sea-captains and ship owners had their homes there, but now sheep-rearing and fishing are the main occupations.

Olib, directly east of Silba, is almost twice the size. It too was leased out by the Venetians. By the sea, in Banve cove, are some Roman foundations, the ruins of a monastery, and the church of Sv. Pavao which was deserted and destroyed around the year 1200. There are no surface streams or springs for water but wide grazing areas for sheep. The farmers make a habit of taking their flocks to the neighbouring small islets off-shore to take advantage of every scrap of grass.

To reach Dugi Otok from these northerly islands one passes Molat. Molat is long, thin, and shaped like an inverted Y, its arms enclosing a very large and sheltered bay. It has three villages in the interior, and sea-level hamlets provide landing-stages for them. Sheep, fishing, and farming are the present-day props of the economy but, like Silba, it was owned by a Zadar Benedictine monastery in the middle of the eleventh century. Of more recent and bitter memory, the fascist Italians established a concentration camp, which became notorious, during their occupation of 1941 – 3.

Dugi Otok, meaning Long Island, is the biggest in the group and has a population of 5,500. It has an excellent natural harbour at Božava. Božava is one of those perfect holiday spots with gently sloping wooded hills and some fishermen's cottages near the quay. A new hotel has been built which will bring badly needed help to the local economy. There are still about 200 houses for only 150 people!

The southern part of Dugi Otok is a typical barren Karst upland, and most of the population live on the side facing the inner islands of Ugljan and Pašman and the mainland. There are five villages of which the main one is Sali. The island was inhabited early, and there are archaeological remains from the New Stone Age and traces of both Illyrian and Roman life. On one neck of land called Mala Proversa a large Roman country house has been partly preserved, and the traces of the sewage system can still be seen in the sea. The mosaic floor from the house is now in the museum in Zadar.

Sali has just over 1,000 inhabitants. Its industries include a fish-cannery, a salt fish and fish meal plant, and a small boatyard for the construction of wooden craft. The name Sali is derived from the nearby salt-pans and was first recorded in 1105. Except for a narrow road between Sali and Luka, further north-west, there are only donkey tracks on the island.

Iž is the only island in north Dalmatia where pottery is made. The potters produce beautiful and simple pots and bowls with good symmetry — honest working pots and flat bowls for fruit — and a few beautifully decorated for ornamental use. Another cottage industry exclusive to Iž is the making of hand-worked cloths and rugs which are made in the homes and organized by one of the women. The designs are in tapestry cross-stitch — roses or Greek urns, or local peasant motifs. Iž women are not only full of energy and life, they are also extremely pretty. They nearly all wear black — so much so that, when President Tito visited them, he asked why so many were in black costume with no colour to relieve it. The people conducting his visit glossed the matter over, saying it was the local custom, but Tito was not satisfied and asked the women themselves. The real reason is that they lost so many husbands and sons in the war as partisans; and once a widow goes into black, she wears it for the rest of her life in a peasant or island community.

Iž lies directly between Dugi Otok and Ugljan and has ten small satellite islands. The first tourist hotel is going up near the village of Iž Veli, but the island is still in an untouched and unspoilt state. Iž grows vegetables and various fruits and also supports 87,000 olive trees. It has been inhabited since Illyrian times, and Constantine Porphyrogenitus called it Ez. The Slavs arrived in the eleventh century, and some élite Zadar families with Venetian-sounding names such as Canagietti and Fanfogna strewed their palaces around in the Middle Ages. Near Iž Mali a round eleventh-century chapel still stands, and there is a fortified tower near by. The island lies in one of the perfect Adriatic positions, surrounded by calm water. Yet it is close enough to be linked with Zadar and mainland sophistication by a daily passenger boat.

I said goodbye regretfully to this Zadar group of islands. In spite of their closeness to, or perhaps because of, Zadar, their individual island character does not come through so strongly as that of other islands. They seem, by and large, palely mirrored rather than independent, though, in contrast, the young men who go to the mainland and progress, pick up a certain toughness and 'rough fur' which you can take or leave; yet this, too, seems to be only superficial and not an inbred quality.

Zadar

Zadar, on the other hand, has through the ages elbowed sharply for freedom. The atmosphere crackles with self-determination and the pride that comes after centuries of never submitting without a fight. Even after the last war when Yugoslavia was a firmly united country under President Tito, Zadar youth showed its feelings during the Trieste crisis by defacing the Lions of Venice which had for so long decorated the walls of their town.

Zadar is full of archaeological treasure, overshadowed, perhaps, by the vast circular church of Sv. Donat, with its huge blind-arch walls built probably by Bishop Donatus in the early ninth century. It stands on the site and remains of the large Roman forum. The fascinating thing is that, in spite of its great height and size, the church was built without foundations. Huge round blocks of stone, taken from Roman columns, were used indiscrimately in its walls at ground level, some of them sideways on, so that one marvels that the building could have stood so long on such precariously perched round bases. Other Roman fragments and inscriptions were used as wall stones, some of them upside down as if the builders wished to exorcize the influence of such pagan thoughts.

There are still massive walls round the original area of Zadar, which was built on a peninsula in a defensive position so near perfect that it seems almost too good to be true. You have only to wander about inside the walls to come across unheeded Roman columns, architraves, Venetian gateways and numerous ecclesiastical treasures in spite of the fact that 80 per cent of the buildings were destroyed by bombing during the last war. All the churches are worth visiting, and one point of particular interest to English visitors is that the upper stages of the fifteenth-century cathedral bell-tower were constructed by T. G. Jackson in 1892; he modelled it on the cathedral bell-tower of Rab. It was the same Sir Thomas Jackson who wrote so attractively about the Adriatic in the nineteenth century. But the church which, besides the outstanding Sv. Donat's, remains in the memory is that of Sv. Šimun which stands near the former Governor's Palace. The church itself is overloaded with the Baroque influence, but what makes it outstanding is the mummified body and the sacrophagus of St. Simeon. The silver coffin, from the fourteenth century, weighing 250 kilogrammes, is decorated in great detail and shows scenes from the life of St. Simeon and, also in relief, fascinating pictorial records of the history of the time. There is the entry of the Hungaro-Croatian king Ludovik into Zadar; the *Ban*, Pavao Šubić and his wife; Queen Elizabeth, daughter of the Bosnian *Ban* Stjepan Kotromanić; and the death of her father. The coffin was created by goldsmiths and silversmiths from Zagreb, Milan and Zadar between 1377 and 1380, and was commissioned by Queen Elizabeth who was Ludovic I's wife. In the coffin there is a Gothic crown and a frail little pall embroidered with pearls covering the brownish remains of the saint, who looks pathetically small. To study the pictures on his coffin is to look through a window at medieval Croatian history. For a wider sweep of the past there are numerous museums and collections in Zadar.

Today Zadar is a busy port with passenger and cargo ships coming and going night and day. It is an important tourist centre, and has a yacht marina with boats to be hired, berthed or repaired. The town is a higher education centre for the area, with several faculties. It is also an industrial centre which includes precision engineering, textile and food industries, shipbuilding, a cigarette factory, and a distillery which produces the famous maraška cherry brandy and maraschino.

It is busy, forthright, and seems young; yet it is a very old city indeed. It was inhabited in the Neolithic Age. After Illyrians and Greeks, the Romans came with great civilization, calling it Jader or Jadera. It became a Roman Municipium in 59 B.C. The Romans expanded the original Liburnian and Greek settlement and laid out a regular street pattern with longitudinal and transverse streets, most of which are still there under the present-day surface. They built the huge forum, 90 × 45 metres, and a basilica; also a temple to Jupiter, Juno and Mars with a portico, the remains of which are still visible. The Croatian Slavs arrived in the seventh century. We know that they came first as foot soldiers with the warlike Avars in the sixth century. The Avars, coming from the

Danube, raided, plundered and devastated and then disappeared again. The Slav Croatians were a more gentle and peaceloving people, slow to anger unless exasperated like most Croats of today, but not easily coerced.

In the ninth century Zadar came under Frankish rule but, under the Treaty of Aachen in 812, was returned to Byzantium. Then, in the tenth and eleventh centuries it was ruled by the Croatian kings of Nin. The tenth to fifteenth centuries were known as Zadar's 'Golden Age'; during most of this time it had autonomy. In 1105 it recognized the claim of the first Hungaro-Croatian king, Koloman, and from that time on there was often conflict with Venice. By the Middle Ages Slav nature was asserting itself. In 1177, Pope Alexander III visited Zadar (it must have been the same journey on which he consecrated the cathedral of Rab). The Pope rode from the harbour in Zadar on a white horse as far as the cathedral of St. Anastasia (Sv. Stošija) and, according to the chroniclers of the papal voyage, the priests and people of Zadar emerged to meet him singing songs in the Slav language: *immensis laudibus et canticis in eorum sclavica lingua.* A slightly defiant welcome if one considers that Rome had been trying ruthlessly to stamp out the use of the Slav language in the churches for about 250 years!

At about the same time, the guilds or brotherhoods began to appear. Zadar, always enlightened and always ready to try something new, was the first city to introduce professional/religious brotherhoods into Dalmatia. The Goldsmiths' guild was in existence in 1176. Altogether there were no fewer than sixty-five societies. All members had equal rights and duties, and the aims were educational, religious, humanitarian, cultural and political. Two of them were granted the privileges of being able to obtain the freedom of one convict. This privilege, unique in Dalmatia, was granted by the state. Some of the brotherhoods even admitted women members. The guilds owned considerable land around the surrounding villages and allowed money loans at the rate of 6 per cent with a security on houses and property. The religious brotherhoods took on social duties which included fire-fighting, the burial of plague victims, and even the provision of dowries for poor girls. The lay brotherhoods, like similar ones in Britain, felt it their duty to watch over the rights and advancement of the crafts and trades, and were also the champions of the ordinary people against any attempt at domination and encroachment by privileged and ennobled families. Zadar was aware, even in the Middle Ages, of the necessity of guarding human rights.

From 1202 Zadar felt the back-lash of European treachery. The city was a sore point between Venice and Hungary, and in order to bring Zadar to heel Venice managed to get the Fourth Crusade to transfer its attentions to Zadar before going east to capture Constantinople in 1204. The Crusaders, arriving in Venice from France, were to be transported to Palestine in Venetian ships, but they were short of money, and Venice persuaded them to attack and subdue Zadar (a Christian city) by offering to defer the payment of the 34,000 silver marks due for transport. Under the Treaty of Zadar in 1358 the city came under the control of the Hungaro-Croatian king, Ludovik I. After Ludovik's death Zadar recognized first the sovereignty of the king and emperor Sigismund of Luxembourg, then Ladislas of Naples. This man, a member of the house of Anjou, in 1403, was crowned in Zadar 'King of Hungary and Croatia'; it was he who sold his Dalmatian rights to Venice in 1409 for 100,000 ducats. After this Venice, taking no chances, restricted much of Zadar's political and economic autonomy.

After the fall of Venice in 1797 Zadar, like the rest of Dalmatia, was ceded to Austria and a change of soldiers and administrators took over in the game of human chess. The Austrians were beaten at Austerlitz and Dalmatia was ceded to France by the Treaty of Pressburg in 1805. The French arrival provided a welcome change. They opened a *lycée* and under the direction of the humane Marshal Marmont improved the physical state of the 'Illyrian Province' as Napoleon called it. Roads were built, commerce improved, agriculture given a helping hand and piracy put down. Zadar remained the capital of Dalmatia. The first Croatian newspaper appeared in Zadar during this time. It was called *Kraljski Dalmatin* — the Royal Dalmatian — and was published between 1806 and 1810. It was to be over 100 years, until after the Second World War, before Zadar was freed from foreign domination. Before this final liberation, the Allies bombarded the city many times between 1943 and 1944. In one raid alone (28th November 1943) 1,500 people were killed in fifteen minutes, while on 9th October 1944 the Allies attacked seven times.

All through history Zadar has had to pick up the pieces of adversity. When the Turks came between 1573 and 1648 the population dropped from 60,000 people in 280 villages — the figures quoted in a Venetian document of 1524 — to 5,500 as the inhabitants escaped to the islands. In 1921, when the Italians took over, 7,458 people left Zadar and were replaced by Italian immigrants. When the city was flattened, it became a challenge to them. Now 80 per cent of the town is modern chromium and glass. It is also very prosperous. The women sell their produce in the vegetable and fruit market, ignoring the stones and columns of history around them. The huge round church of St. Donat's and the archaeological exhibits arranged in its upper gallery echo to concerts of chamber and church music, the acoustics being peculiarly good. At one side of the forum, the modern Hotel Forum bustles with local people as well as visitors, and the modern pillars of its pavement café blend curiously well with the remains of the Roman forum at its feet. Zadar promotes energy and bustle; it always has done!

The second and southerly group of North Dalmatian islands, numbering about 40, are linked with the mainland town of Šibenik. This does not include the offshore Kornat islands which amount to a further 125. Among the more important are Murter, Žirje, Zlarin, Prvić, and Krapanj. Šibenik is built at the mouth of the great Krka river. Like the Zadar group, the bigger islands are economically dependent on Šibenik but, unlike the Zadar group, they have developed through the centuries a character of their own.

Zlarin is 40 minutes away by small steamer from Šibenik, including the journey along the narrow passage which protects the town from the open sea. This narrow deep-sea cutting has a tourist camp on one side and submarine pens cut into the rock at sea-level on the other. The passage has to be negotiated by bigger ships with great care and at slow speed, and the captains must wait for an 'all clear' signal from the beacon on the ancient fortress of St. Ann above Šibenik before they may proceed.

The boat ticket to Zlarin is less than 20 np. The island (8·2 square kilometres) has two ridges with a fertile valley between, and the port of Zlarin is the only village, apart from one small hamlet called Borovica. There are about 1,000 islanders. It is a broad, delightful port with houses of cream-coloured stone set well back from the sea on both sides of the harbour. The colours are pale and clean, and the sky is wide. Zlarin women wear their own island dress every day. The black pleated skirts are worn with white, wide-sleeved blouses and embroidered and braided bodices; sashes and kerchiefs are coral red, which is not surprising because, on this small island, coral has been among its industries for centuries. Zlarin has always had a lively export of sailors. Zlarin women are extremely pretty. They also have a big reputation for fidelity and chastity, which is just as well for the peace of mind of their sea-going husbands.

Zlarin is a friendly and sleepy little island, charming for a day trip or a week's stay. Much of the local life is carried on with the aid of small boats. A boat will bring the grapes at harvest time to the quay; a family's fish for lunch will be caught by the family and brought back to be cooked; needs for the village arrive by boat, are dumped on the quay and sorted out by the women. Zlarin is always busy!

The *Koralja*, as the small coral workshop is called, is up a side street between the houses. In the upper rooms, women in island dress work the natural coral into rounds and ovals and torpedo-shaped pieces; they pierce them, polish them and grade them, all by hand. The only mechanical aids are small hand-turned grinding-wheels and metal-piercing drills which are turned by the string of what looks like a home-made bow. The workers hold the half-inch pieces of coral with their fingers against the grindstone. There is coral dust everywhere, but the pieces they polish have a rare depth of colour and go to make attractive rings, bracelets, and necklaces. It is said that Adriatic coral has a deeper, better colour than any other. One old man in Zlarin, a retired schoolmaster, has a book indicating the best coral beds, written in copper-plate manuscript by one of his forebears; the industry has been going for 200 years. In 1937 the *Koralja* got a silver medal for its work in the Paris International Exhibition of Arts and

Techniques. The raw coral is kept in a huge iron chest with a lock composed of eighteen separate catches. The mechanism takes up the entire lid space — about 1 metre by half a metre — and is a prize exhibit on its own. In the face of modern commerce and factory industry, one hopes that this little oasis of handicrafts will be able to keep going in a changing world.

Zlarin lies on the southerly side of the sea entrance to Šibenik; on the other side is Prvić, smaller, but with two villages. Prvić port is situated in a wide sheltered bay. In the evening light, the village takes on a sort of Canaletto light. The shore is lined with solid stone fishermen's houses. The church (of the fifteenth century) has some Glagolitic works, and until the war the Glagolitic text was still used for the services. The villagers, between six and seven hundred of them, are waiting for water, and expect it in 1971. It will be brought under the sea from Šibenik. There is an air of waiting generally on this small island; a pause as if the struggle of island life has become too difficult, and everything now awaits the arrival of water and tourism. Prvić Luka is a big village. Records indicate that in 1603 there were 150 inhabitants and by 1775 this number had risen to 971 as people moved from the mainland to escape the Turks. The Franciscan monastery, on a hill and now unoccupied, was founded in 1461.

The second village, Šepurine, over the ridge on a second wide bay, has a wide, sweeping shore and they plan to build a tourist hotel.

The island, of course, is very small; but it has a variety of trees including Phillyrea, *quercus ilex* (holm oak), arbutus, juniper and terebinth. Before the devastation of phylloxera it was a good wine centre. The new imported vine stocks bear less heavily than the original variety, so that more land is needed to make a profit in grapes and many of the vineyards are now derelict. The young people go daily into Šibenik to work — there is an excellent boat service — but it is sad to see the neglect of vineyards, even if they are not profitable, and a little disturbing when a community seems to be staking its whole future on the industry of tourism. Tourism can be fickle at the best of times, and if an island loses its own individuality, its peasant economy and its way of life in the gamble, it has to remember that there might not be enough interest and island culture left for tourism to feed on. The industry of tourism can only thrive if it is complemented by the solidity of native life.

Murter, a third island, is further north-west and can, in fact, be reached by car or bus from the Magistrala Highway,[1] there being seven buses a day from Šibenik. Like its sisters, Murter has the same clear-washed watercolour look about it. But just as the nature of one sister can differ from the others, so Murter has a delicious heady air, like the favourite in a family. Tourism has come to Murter, but has not crowded it, and the island has kept its agricultural and shipbuilding economy alive. There are four villages, the first of which spreads over both sides of the low bridge, only 2 or 3 metres above the water, connecting the island to the mainland. This is Tijesno, and the bridge, which can be raised to allow the passage of boats, is 37½ metres long. The village is quite large with a population of farmers and fishermen numbering about 1,500. The road leads from Tijesno both south-east to Jezera, a quiet fishing port, and north-east to Betina and Murter village. Only tracks cross the south-eastern end of the island. The south-west coast, facing away from the mainland, is predominantly steep, unlike the other side of the island, and is indented with beautiful small coves. One, Koširina, is a useful anchorage for small craft when the *Bura* is blowing. Adriatic seamen are too knowledgeable to treat this wind with anything but the greatest respect!

On the other side of Tijesno the road follows the shore to Betina, an attractive village set on a small point. Betina, with a population of 1,000, supports a shipyard which constructs small ships and yachts. It is not a very old settlement, though on a hill near by stood the Illyrian-Roman town of Collentum. Betina was founded by the inhabitants of Vrana when they escaped from the Turks in the sixteenth century.

Murter village in the north-western corner of the island has about 2,000 inhabitants. It is built on the steep hill of Vršina, and on its highest point there is a small church, Sv. Rok. Once at the top and in the square yard outside the church, the air has a heady

[1] The new highway which runs from end to end of the Yugoslav Adriatic coast.

upland tang to it and the view of the surrounding islands and the sea is fantastic. All the Kornat islands are spread out below. Their bare backs and uninhabited sides look like so many basking whales, or perhaps like drops of water splashed on to a dark crystal-blue polished surface.

The road from Murter village goes inland to cross an arm of the island and arrives at Slanica which faces out to sea towards the Kornats. Slanica is in a beautiful bay edged with pine trees, and is a complete tourist village with a good hotel and restaurant, a camp which provides small cabin-like cottages, and wonderful bathing.

Much further out to sea, and almost forming a tail to the Kornat islands, is Žirje. It is the largest of the inhabited islands in the outer chain, covers 15·43 square kilometres and lies 22 kilometres south-west of Šibenik. The only village of any size is called after the island and has a population of about 500. It is very isolated and quiet, except on Sundays a small Jadrolinija ship makes the return journey every day between Žirje and Šibenik.

The Kornat islands, like the other islands in this area, show traces of Neolithic occupation, Illyrian earthworks, dry-stone walling, and stone tumuli. On the island actually called Kornat an early Christian church with a semicircular apse has a smaller chapel built inside it. There is a Roman votive plaque built into the doorway. In the early Middle Ages a large fortress was built on the top of a hill — a good defensive forward observation post for the protection of Šibenik! Otherwise the whole range of the Kornati, except for a few fishing cottages, is bare. No one lives there permanently, and the islands have no fresh water except for a few shallow patches after rain. On the island of Žut, there are some holm-oaks and a few arbutus trees. In the depressions of the karst rock there is a little red soil where olive trees and grape vines grow. Most of the land is owned either by the inhabitants of Murter or the families in Sali on Dugi Otok. The sea round these almost fabulous isles is rich in fish, and there is some sponge-diving. With no regular boat services to the islands, people with their own small boats regard them as a semi-paradise. There is one camping site on Ravni Žakan.

The name 'Kornati' may have come from the Latin word *corrimare*, to crush, as if some god had crushed the rock and then thrown the pieces into the sea. But in fact the islands are parallel with the great Dinaric Range on the mainland, and are the mountain-tops of a second ridge showing above the water. They occupy, between them, 320 square kilometres, and while most of them are of chalk limestone, the island of Žut — which means yellow — and the cluster of smaller islands near by are made of shell limestone. The scattered cottages on Žut are empty most of the year, but farmers on Murter bring some of their sheep to winter there.

The island called Krapanj, south-east of Šibenik and within 300 metres of the mainland, is also one of the North Dalmatian Group, but very different from the Kornats. The one village faces the mainland and, because the island is so flat (only 7 metres at its highest point), seems to cover the entire island, though in fact there is also a considerable area of open maquis and grass. The village has a most attractive medieval air about it. During the Middle Ages Krapanj belonged to the Šibenik chapter, and the Franciscans, arriving in 1446, built a monastery with Renaissance cloisters, completing the church in 1523. The refectory contains a picture of the Last Supper by Francesco da Santacroce. Krapanj must have been teeming with life in those days. By the beginning of the sixteenth century the Turks were encroaching almost to the coastline, and the inhabitants of the near villages took refuge on Krapanj, small as it is. Today there is a population of about 1,600, and agriculture is attended to mostly by the women of the island. By tradition, the men are sponge- and coral-divers. The village supports a sponge refinery and, as on Zlarin, a small coral workshop.

Still further south-east is Primošten, an even smaller island but beloved by holiday-makers. It is now joined to the mainland by a causeway. Although so small and long celebrated as a tourist resort, Primošten has taken care to keep its character as an island village and produces some of the best wine in the area.

It was originally called Bosiljine and was an island with protective walls and a drawbridge. It was developed by eighteen families who came from Bosnia when the Turks drove them out. The families went to Šibenik in 1571 and settled on Primošten. They tended flocks of sheep and tilled the fields on the mainland, but each evening, when a tattoo was called on a drum, they repaired behind the drawbridge to the safety of the small island at night. In the Šibenik museum there is a map, about 300 years old, which shows the island before the causeway was formed. Gradually the sea brought silt to make a narrow 'bridge'; the people reinforced it with stone, and it was after this time — about 270 years ago — that the name Primošten was evolved from the verb *primostiti*, to bridge. The defence walls and towers were pulled down at the end of the last century (during the Victorian era the whole of Europe seems to have been imbued with the desire to 'let in more light' and tear down old structures, however historic!). In spite of its tourist prosperity, the spirit of the old settlement is still very strong. In October, the heady smell of wine-making exudes from each cottage and the women walk purposefully about with baskets of grapes on their head. The island dress is very gay, with large bright satin aprons of various colours. Primošten was bombed during the war by the Italians and the church at the top of the hill was badly damaged, the east end with the altar being destroyed. Now the Italians have rebuilt the damaged portion in very modern style. There are tall slim arches beyond the altar, in the form of an apse with hidden lighting, and a modern altar of plain white stone, supported on two black isosceles triangles is starkly impressive. This blending of ancient and modern stonework into harmony is striking and moving — a perfect example of the joining of extremes.

The friendly atmosphere of Primošten seems to have inspired others. In the garden of the Adriatic Hotel, built near the old town on a second wooded point, there are twenty-eight small oblong flower-beds. They were planted in 1962 by the delegates, who came from all parts of the world, to an Esperanto Summer Camp arranged by a group of Zagreb astronomers. Each delegate made a small garden, incorporating in it a little of his own country. The Japanese brought soil from Hiroshima, the Hungarians burnt the paper they had with them from Hungary and put it into the soil, and so on. The new hotel has a seawater pool equipped with a domed, three-leaved glass roof which allows the ultra-violet rays to penetrate; electro-physiotherapy appliances; masseurs; Turkish baths; and various outdoor sports.

Šibenik

Šibenik, 26 kilometres away, is the most important architectural centre in the area. Unlike most of the Yugoslav Adriatic towns which had their inception much earlier, Šibenik was originated by the Croats early in the Middle Ages. The town is well protected from the sea by its narrow approach channel which then opens up into a huge inland natural harbour, the powerful Krka river sweeping in from the mountainous land behind. The medieval town is perched on the side of a steep hill above the harbour. It was settled in the tenth century as a Croatian tribal fortified village and is first clearly mentioned on Christmas Day, 1066, when King Petar Krešimir, in Šibenik, issued a decree giving royal protection and freedom to the Benedictine establishment of Sv. Marija in Zadar. Throughout its history Šibenik has been lion-hearted and belligerent, championing liberty and taking on all comers, like a small dog fighting a lion. Twice it was burnt to the ground by Venice — in 1116 by the Doge Ordelaf Faliero, and in 1378 by the Venetian admiral, Pisani. Each time the inhabitants set to work to rebuild it. From 1409 to 1412 it resisted Venetian capture, but finally had to surrender. The Venetians stayed for nearly 400 years during which time the Turks encroached almost to the town walls in the 250-year struggle between Eastern and Western influence and religion. The Venetians curtailed and curbed the city, but during the entire period the tremendous vitality of the Croatian citizens of Šibenik managed to keep alive their own Slav arts and culture.

Buildings and treasures are many in Šibenik, crowded into the crooked streets and houses of this little old town which clings to the hillside. Pride of place must be given to the glorious small cathedral for which Juraj Dalmatinac is responsible. Only by seeing can one appreciate the love of natural things he has instilled in it: the gently wind-ruffled leaves of his stone acanthus; his fondness for children which shows in the delicious font held up by three cherubs in stone, two of them working with earnest expressions on their faces while the third rests his back and giggles cherubically. The stonemason's sense of humour and understanding of human nature is ever present, and shown vividly in the stone faces round the outside walls of the cathedral, among which one can tell his friends, his enemies, and those who owed him money! The cathedral with its splendid barrel roof speaks like a sonnet, while the ancient part of the town tells a story like an epic ballad. Šibenik's medieval skirts sheltered the hero-pirates whose exploits gave the townspeople something to laugh about during the dark days of foreign oppression. The present century has brought hitherto undreamed of development. The population has swelled, from 17,000 before the war to 30,000, and Šibenik is now an important Adriatic port and industrial centre. Aluminium works, electrode and ferro-alloy plants, and mechanized harbour installations, together with technical schools and other higher education centres have brought a bright new prosperity to the area.

There is one other small island near Šibenik which makes a fascinating excursion; inland, the island monastery of Visovac is surrounded by the swirling waters of the Krka river. To reach it entails a bus or car journey and a trip by small motor-boat. This alone is worth doing. The road from Šibenik towards the river finally shoots down a steeply hairpinned slope from the uplands to the river's edge. From the top, there is a magnificent view of three broad channels of water — the Čikola river to the right, which joins the Krka before one's eyes, and the joint flow towards the sea on the left. The water, in the wide sweep where the rivers join, is quite a different blue from the sea, with dark emerald patches in the shallows. The rivers are full of fish and our boatman, while he waited for us, caught two fat trout and cooked them on the river bank for *marenda.* The river Krka is very broad, almost like a lake, on the way to Visovac. The reflections are fantastic — complete mirror images of the shores with their willow trees, their minute deserted gravel beaches, and small overhanging rocks. Dragonflies, huge lacy creatures, flit erratically by, and the birds, encouraged by the noise of the boat engine, sing louder. The monastery island is about 6 kilometres upstream and lies in the middle of the river which is about 700 metres wide at this point. The Augustinians founded the monastery when the island was known as Lapis Alba (White Rock) and a little of that first building is left; a few pillars built into a wall round a patio, the well, and the original church which is now used as the sacristy. It was dedicated to St. Paul, and the island was called St. Paul until the seventeenth century. The Augustinians left in about 1440 and, it is believed, removed their foundation to a small island near Korčula. About the year 1445 some Franciscan monks, driven out of Bosnia by the Turkish invaders, moved to Visovac. The Turks caught up with them in 1522, but the monks were able to make a bargain with them, agreeing to pay tribute. They stayed on Visovac until 1648. But the Franciscans actively supported the insurgents of the Cretan War (Venice ruled Crete until 1645, when the Turks gained possession and dominated it until 1898). The unrest in Crete continued from 1645 to 1669 before it was finally crushed by the Turks. The Franciscans, fearing reprisals, left the monastery. In 1674, after the Cretan struggle was over, they applied to the Sultan Mehmed IV for permission to return. In the Refectory the document giving them permission still hangs. It is a lovely piece of work and looks more like a beautiful and futuristic picture than a legal document signed by a monarch. The characters, in black, gold, pink and blue, have a flexible symmetry, and the sultan's signature at the top is an intricate pyramid of flowing line.

The landing-stage for the island is a wooden platform overshadowed by tall trees. A path leads to the monastery through the large gardens. The present church, which was enlarged in 1694, is interesting for its two main altars, of equal status, in the two Gothic apses of the Franciscan church. On the north side of the monastery you can peer down the original well dating from the Augustinians' time. It has a crenellated wall and a solid little roof structure with red tiles, supported on four pillars — a stone baldachin with the charming air of a medieval folly! The Franciscan monks have always served as parish priests to the area, and in 1568 a novitiate was opened. The entire island has a happy, young atmosphere about it. The library contains a small and valuable collection of books including the incunabula of 1490 printed in Venice, some very early handwritten works and one of the largest collections of Turkish documents in Croatia, some 620 in all. The most famous and valued book is *Aesopi Fabulae, Brixiae, by Boninus de Boninis de Ragusia,* and dated *MCCCCLXXXVI.* This attractive and beautifully illustrated work was drawn and printed by Dobriša Dobričević, who was born in the small island of Lastovo (see Chapter 10).

The Brothers live entirely on what they produce. Besides flowers and shrubs, the garden yields fruit, nuts and vegetables. Livestock includes pigs, chickens, and even peacocks, and there are magnificant fish in the great river which washes both sides of the island and is gradually enlarging it by silt deposits, so that it is now considerably bigger than the original 'white rock' of Augustinian days.

You cannot learn much about Dalmatia without coming across the work of one of their greatest sculptors, Ivan Meštrović, whose forceful figures have a reputation throughout the artistic world. His statues and groups are famous, modern and vivid. One of the best known is his statue of Bishop Grgur of Nin outside the Golden Gate in Split; another is his figure of the stonemason Juraj Dalmatinac which stands by the cathedral of Šibenik. Meštrović (1883 – 1962) was born in a poor district of the Dalmatian hinterland. When he was four years old he started to carve wood with his pen-knife, and when he was fifteen he produced a remarkable stone sculpture of a 'Bosnian on Horseback'. Later, though poor, money was collected to send him to the Academy of Fine Arts in Vienna. Throughout his life he poured out strong and arresting sculptures. A number of his early works can be seen in the church of Sv. Ante in Drniš. Eight kilometres away, at Otavice, is his mausoleum. It lies over the mountains, about 30 kilometres north-east of Šibenik.

Chapter 5 The islands of Čiovo, Trogir and Nin

South-eastwards along the coast beyond Šibenik there is a sudden dearth of off-shore islands to shelter the coast until one approaches Trogir in the vicinity of Split.

The first inhabited islands in this section are near the town of Trogir and are both called Drvenik — one Mali and one Veli — and they are invisible from the Magistrala Highway as, at this point, the road goes inland to skirt the port of Marina and its minute medieval village. This is an enchanting corner, its slim bay lined with profusely flowering oleanders.

The name 'Drvenik' crops up several times along the coast, and is a derivation from the word *drvo* meaning wood. The two islands are very small, the first not much over 3 square kilometres and the bigger, Veli Drvenik, about 12. The former has a population of 280 and the latter just over 700. A weekday passenger boat from Split takes two and a half hours to arrive, making eight ports of call. The islands are well cultivated — 'away from it all' yet close enough to the cultural centres of Trogir and Split — and offer the ultimate in peace, quiet, and the simple life.

Čiovo

Not much further east, and almost touching the mainland in front of Trogir, is Čiovo. Unlike the Drveniks, Čiovo has very poor soil; most of the island is covered in maquis shrubs and heather, and the ground is rock and stone. It does, however, manage to produce figs, almonds and grapes and, in a few sheltered spots, some citrus fruit. The six villages have a combined population of around 4,000, but life is very much geared to Trogir. Čiovo, which is attached by bridge to Trogir has several satellite islands off its south-west coast adding interest for the lazer, bather and boater.

In classical times it was known as Bua, Boa and various other names, Bua meaning 'a place of snakes', though there is not much evidence of this now. Indeed, one theory is that a certain cattle-consuming constrictor once roamed about there; rather like Loch Ness and its monster! The island was inhabited in Roman times and became, during the latter part of their era, a place of political exile. Later, it was almost deserted, except for a few hermits. But in the fifteenth century, while Turkish attacks were becoming fiercer on the mainland, refugees settled in the island's interior, and even the people of nearby Trogir — itself an island, technically — moved over to Čiovo. As with the island of Pag, Čiovo too was divided. The eastern part belonged to Split while the western area belonged to the commune of Trogir. Overshadowed though it is, artistically, by the riches of Trogir, Čiovo does have a few treasures. There is an early Romanesque chapel, Gospa Kraj Mora — Our Lady Beside the Sea — with a fifteenth-century wooden polyptych. In the village of Sv. Križ (Holy Cross) there is the church and Benedictine monastery built by a local master mason of the fifteenth century, Ivan Brahanović, with Gothic paintings on the Chapter House ceiling. As early as 1214, a Franciscan order lived on Čiovo. An early Romanesque church at the village of Žedno, inland, is worth seeing, and on the cliff of the southern coast stands a chapel dedicated to the Virgin of Prizidnica, with a painted Gothic crucifix. Also belonging to this little chapel is a Byzantine icon, but this is now in Slatine. Minor roads fan out from the bridge connecting Čiovo with Trogir: to Slatine; along the crest of the central heights to Žedno, stopping short of the highest peak, Rudine (218 metres); and west as far as Donji Okrug. Apart from these, only tracks cross the island.

Trogir

Trogir is one of those rare places which seem to touch the heart even before the eye has had time to perceive. Its outskirts, and even its attractive vegetable market, sprawl across the narrow strip of seawater — more like a river than a sea — which separates the old town from the mainland. Indeed, it is hard to believe that Trogir is built on an island, so narrow are the waters between, though this fact has had its defensive uses during the course of history.

Trogir is 27 kilometres from Split and is situated at the western end of the broadly curving Kaštela bay while Split is at the eastern end.

At Trogir, across the bridge over the seawater channel, one faces the walls and gateways of a medieval world. Outside the walls a cooper works at his trade; the new, curved wood of wine barrels gleams darkly within his workshop. Close by, there is an enormous wickerwork cage, twice as high as a man, in which an assortment of birds live together, including jackdaws, fed by the local population. The entire scene has the flavour of an etching from the middle centuries and puts one in the mood for what is to come. Through the north gate there is a maze of alleyways, leaned over by ancient stone façades; Gothic windows, with their stone sills supported by lion-heads; carved stone guttering and corbelling, and the remains of elegant Venetian balconies. Here and there the streets open out into small squares or courtyards, and the sight of a man tinkering with a car engine or a tourist strolling by in shorts seems incongruous. The people of Trogir are closely knit and essentially friendly; ask who is selling good wine and they will pass you from hand to hand, or break off their own business to accompany you. During the winemaking season I have sat at an open-air café and watched countless women pad by carrying pitchers of fresh juice on their heads for themselves or neighbours to make into wine — mutual aid and a new slant on the word 'communism' which has nothing to do with politics. None of this has changed for centuries.

Among the beautiful old places on the way to the main square is the baroque Garagnin-Fanfogna home, now the municipal museum. Once in the main square — the Narodni Trg — the cathedral claims one's immediate attention. Solid and massive, but with flights of marble fantasy, it has blind-arcaded walls and other Romanesque decoration. There are billowing rounded apses, and beautiful and intimate murals, lovingly and painstakingly carved by masters of stone. The square bell-tower was built in three different sections and three different styles, between 1422 and the sixteenth century. The simple Romanesque rounded portal of the cathedral facing into the square is dated 1123, the date when building began. But it is the portico in front of the main doors which draws one towards the church. Inside this square outer chamber is the famous portal carved by Radovan and his colleagues in 1240, and known as the most important example of Romanesque art in Dalmatia. It is tremendously complex, with each small section perfect and detailed. On the outer sides are two lions supporting the figures of Adam and Eve, who modestly hold their own fig-leaves in place. The outer pillars supporting the three-tiered arch are decorated with small saintly figures, one above the other and each with a different face. The central pillars are carved with scenes depicting the months and the seasons, while the inner, rounded columns have hunting scenes and fantastic animals and thick foliage. Each scene is tiny, and the carving so deep that there is space behind, like the intricate wood-carving of Macedonia. At the very base of the columns Saracens and other infidels hold up the rest of the porch, in everlasting punishment for destroying churches. Above the doors are other biblical scenes while, inside, the cathedral is rich in works of art.

Facing the cathedral on the southern side of the square is the city loggia, pillared and open-sided. In its recess there is a wall-relief of 'Justice' of 1471, and the right-hand column has a particularly beautiful Corinthian capital. Behind the loggia is the dark medieval chapel of St. Barbara (ninth and tenth centuries), simple and innocent — a lovely example of early Croatian church architecture, while on the west side of the square is the headquarters of the tourist office, housed in what was the Cipiko palace. In the entrance hall a large carved wooden cockerel, painted black and red, juts out aggressively high up on the wall. It is the figurehead from a Turkish ship, and the remaining one of two which were kept here as mementoes of the Battle of Lepanto in 1571. The cockerel was captured from the Turks in that famous confrontation when Ottoman naval power was finally crushed south of the island of Corfu.

A street at the far end of the main square leads towards the sea and the 'Town Gate'. Near the gate is a Benedictine convent with the church of Sv. Nikola. There is a wooden Gothic polyptych and a Gothic crucifix in the church dating from about 1400. The convent was founded in 1064 and is one of the oldest in Dalmatia. Although it is a closed order, visitors may ring the bell in order to see one of the most lovely of all ancient Greece's fragments of carving, for it is this convent that houses the glorious relief of the god Kairos. When the door opens one rings an inner bell; through a grating a nun's voice answers. Then a small swivel cupboard is turned from within and a key to the museum lies on the shelf. The room holds some interesting icons and other pictures but, for me, everything pales into insignificance beside the little yellow marble relief which hangs on the wall. God of the Fleeting Moment, that propitious second in time when all is at its best. Kairos holds scales in his outstretched arms. His yellow hair is windblown and his straight-nosed profile has an eager yet hesitant and supplicating expression; a mixture of the ineffable and the infinite. There are plenty of museum copies of Kairos, but none has the power of this exquisite original of about the first century B.C. It was rediscovered in 1928. The convent also houses, in a wall of the cloisters, an earlier Greek tablet recording a decision by public assembly of the fourth or third century B.C.

In 1644[1] another fascinating document was found in Trogir by a lawyer called Marin Statileo who found, in the library of the Cipiko family, a manuscript by the author Petronius: an extract from the novel *Satyricon Libri*. As Director of Pleasures to the court of Nero, Petronius produced this manuscript, 'The Feast of Trimalchio'. The manuscript is now in the Bibliothèque Nationale in Paris; one hopes that it will one day be returned to Trogir.

On the seaward side of the Town Gate there is a second very small but beautiful loggia, built in 1527. When the Venetians arrived in 1420, after bombarding the town for four days to subdue it, they strengthened the fortress, built earlier by the Genoese, and closed the walls round the town. Travellers wishing to enter the city could shelter in the loggia until permission was

[1] I have taken the local date; other sources put it at about 1650.

granted for them to enter. Now, it is a market for fishermen who sell their catch, often tuna, as soon as they have landed it. I bought two small tuna steaks very cheaply and, when they discovered I was English, the purchase was turned into an occasion. Buying there, as in the vegetable market, is an experience packed with politeness, humour, interest (that one is a foreigner) and delight on the part of the vendors at such a felicitious encounter.

Trogir, in the third century B.C., was a Greek settlement with the name of Tragurion. These colonists were of Doric tribe already settled on Vis (ancient Issa). It is said that the Greek name, Tragurion (the town of goats), probably derived from the large goat market which was run by the Illyrian aboriginals. When the Romans came, Tragurium, as it became, prospered as a busy port and was, in fact, the state granary. After the fall of the Western Roman Empire, Tragurium was incorporated into Byzantium between the fifth and ninth centuries. It seems to have been an uneasy time, with raids and forays from marauding pirates and barbarians, and the frequent threat of bigger foes. There was Attila (about 406 – 53). Known as one of history's greatest conquerors and greatest destroyers, he was the 'Flagellum Dei' who galloped across Europe leaving a trail of destruction in his wake. Though he never attacked Dalmatia, it was full of refugees from his path. A few years later in 481, Odoacer took Salona and thenceforth ruled Dalmatia, though neither he, nor his successor, the Ostrogoth king Theodoricus, interfered with the existing social system, leaving the local authorities to carry on their normal work.

Later, in the ninth century, when the Gothic-Byzantine struggle erupted into bloody battle in Dalmatia, the Great Migration across Europe of the Slavs was already in progress. Tragurium, although still paying lip-service to Byzantium, recognized the strength of the new influx, and was already paying tribute — 710 guldens annually — to the Croatian rulers in the ninth century. The dukes of Croatia made their centre near Tragurium, at Biaći. Soon a situation obtained very like the present situation in Yugoslavia. The younger people began taking over the reins of authority from the older Romanish men, recognizing the Croat way of life, just as the younger, and more liberal-minded, men of today are taking over from the old, hard-core partisans who had their doctrine straight from the rigidity of Moscow. Tragurium became Trogir; Tomislav became king[1]; but there were still various factions claiming the title. In the year 1000 Peter Orseolo of Venice, taking advantage of these squabbles, captured some of the coastal towns and islands including Trogir. Svetislav, the jealous brother of the Croat king, was in Trogir, and actually made a deal with the Doge, giving his small son Stjepan as a hostage. Trogir then vacillated between Venice and the Croat kings until the coming of the Hungaro-Croatian rulers at the end of the eleventh century.

A bishopric was founded in Trogir in the eleventh century which was to endure until 1828. In 1107 the Hungaro-Croatian king Koloman granted the city autonomous municipal rights. Sixteen years later, the Saracens overran Trogir and almost destroyed it in their violence. Resilient, it recovered and entered its most prosperous phase in the twelfth and thirteenth centuries. In 1242 it flung a protective cloak round the beleagured Hungaro-Croatian king Bela IV, defending itself behind its rebuilt fortresses and walls. At this time the Mongolian hordes, under the supreme command of Genghis Khan's son Ogodai (who was in Asia) had been rampaging across Europe through Hungary, towards Vienna, and south into Croatia. King Bela had previously taken a stand inside Diocletian's palace at Split. But the people of Split disliked the thought of the Mongol hordes bunched outside their Zlatna Vrata and (though their hospitality to the royal family was impeccable) did not wish to co-operate in battle. Bela was forced to retreat with his court to Trogir.

In Trogir things were going badly, and the Mongols, across the narrow sea channel, called out their terms for his capitulation. Under Bela's orders, the whole population inside the walls of Trogir remained absolutely silent. A devastating situation for would-be conquerors, reminiscent of the anti-climax at a children's party if the blindfolded child calls out triumphantly that he has solved the puzzle, only to find that the rest of the players have

[1] Tomislav was the first Croat to assume the title of 'King' and was crowned at Nin in 924.

trooped into another room for the jellies and cakes. The Tartar's emissary withdrew to take counsel and, due to one of those extraordinary quirks of history, the whole marauding mob then disappeared without further fighting on receipt of the news that their leader, Ogodai, had died. Like homing warheads, they scarcely stopped to pillage except at Kotor, which they burned *en route.* In the meantime, Bela had been taken for safety to a neighbouring island, known even now as Kraljevac, or King's Island.

Dalmatian history, immensely complicated at the best of times, in the thirteenth and fourteenth centuries was even more tangled in Trogir, which found itself drawn into the affairs of Bosnia and Serbia. Trogir elected its overlords from the famous Šubić tribe of Bribir. One of these men, Mlađen III, died in 1348, and the epitaph on his tomb in Trogir Cathedral calls him 'the shield of the Croats'. But Mlađen's sister, Jelena, was the mother of the Bosnian king Stjepan Tvrtko (1353 - 91). He was also by marriage a nephew of King Kotromanić, a Bogomil who later turned Roman Catholic. Tvrtko reaffirmed Trogir's old rights of liberty and self-government and, in return, the people of Trogir supported him with troops in the great and bloody battle against the Turks at the Field of Kosovo — the famous Kosovo Polje — in 1389. So the little town of Trogir took part, far away from the Dalmatian coast, in the most terrible battle of the whole turbulent history of the Balkans.

In 1420 Trogir finally came under the domination of Venice until 1797. Many of the beautiful buildings owe their inception to Venice, and the winged lion of Venice presided over the town. After 1797 Austria seized power until 1918, except for a brief but lively incursion of the Napoleonic French between 1806 and 1814. The French did their best to make up for the lack of Croat culture allowed by Venice. They opened a secondary school, several medical and law courses and a number of primary schools. They also built military roads, while the delicious little 'gloriette' or Belvedere, a neo-classic temple or folly put up in honour of the humane Marshal Marmont, is the perfect, and only, French monument in Dalmatia.

At the end of the First World War Trogir's citizens were, of course, completely unarmed; nevertheless they repulsed a force of armed men who attempted to set up Italian rule in the town. Trogir became a sore point with Italy. In the early thirties some of the young men of Trogir slightly defaced the two lions of St. Mark still extant on the walls of the town. This particular act provoked Mussolini to fury in the Italian senate in 1932. All that actually happened was a chipping of the symbol of Venetian, not Italian, power, dead by then for some 130 years. One can only feel unbelieving astonishment that Mussolini should have been so personally affronted.

Today Trogir is a sunny, peaceable Mediterranean town, with more than a normal share of architectural and artistic treasures. Its enclosing walls have mostly disappeared except for guard towers and a large castle ruin in which they hold film shows. Its people have a 'completeness' of personality — friendly, yet sufficient to themselves; gay, but not fawning; proud, and too proud to overcharge. Turkish coffee is as cheap in Trogir as anywhere along the coast. The open-air restaurants produce inexpensive meals without fuss and with great good humour. Like Dubrovnik, Trogir regarded itself as a kind of city-state in the late twelfth century: a status whose glister, once on the shoulders of the citizens, never quite rubs off in the succeeding centuries.

Nin

Nin, 17 kilometres north-west of Zadar, and almost too small and forgotten to be called a town any more, lies in a lagoon-like sea-water bay. Inhabited since early Neolithic times, with copious Liburnian and Illyrian remains, Nin's zenith occurred during the Roman and early Croatian periods. Now, chickens scratch about among the huge plinths and columns and decorated stones which remain from the Roman forum; cottage walls sport slabs of Romanesque decorated marble; children stare, wide-eyed, at unaccustomed visitors; while farm dogs bark from the tops of hay wains, and donkeys lead their foals, unaccompanied, up the main street.

Nin, for me, has terrific atmosphere. Like Pompeii, it draws me back through the centuries. In Pompeii I have had the feeling that, if I touch one of the stones of the bath-house, I shall disappear into a Roman plane of time. In Nin, near the site of the temple, imagination again takes hold and transportation seems imminent: back to Roman Aenona; back further to Liburnian 'Hemionoi'; back to the Illyrio-Roman cult of the goddess Venus-Anzotica, representing, for the Liburnians, fertility and plenty.

Nin was a civic centre, an island, and a crowning-place for the first Croatian kings. Suddenly it was abandoned by both men and the sea due to the silting up of the Jaruga estuary. But paradoxically, the east Adriatic coast has been submerging at the rate of one millimetre a year, and has sunk about two metres since Roman times, so that prehistoric and some Roman remains have disappeared, while the town, on its island, is now only three metres above sea-level. One day Nin will become a true island again, instead of a hummock of land surrounded by shallow seawater marshes. The small *gostionica* near the bridge at the beginning of the town, simple as it is, provides *pršut*, fish, fresh bread and good wine, while, if you prefer grilled meat, the abattoir is just across the way, and a shout from the *gazdarica* would undoubtedly produce anything you want. When I was there, the bathroom water supply was ample but salt; now, a fresh-water supply has been laid on to the town.

A second chance for prosperity in Nin is that it may be developed as a spa. The Institute for Physical Medicine and Rehabilitation at Zagreb published, in 1968, the results of their investigations into the black and grey silt at Nin which is extremely beneficial as a treatment for some illnesses. It may well be that medico-tourism will be Nin's answer to the future.

Very little archaeological exploration has been carried out in this rich but latent treasure area so far, but fascinating finds must be absolutely crowded just under the present surface of the soil. There is, for instance, the remains of a building on the east side of the forum with rich mosaic floors still in place under a shallow coating of top soil. The decoration includes medallions and vignettes enclosing animal scenes — perhaps the signs of the zodiac, writes Dr. Mate Suić — a fight in an arena and even the names of the gladiators. Whoever built the house must have been a very keen, and maybe superstitious, sportsman, or perhaps a retired and honoured gladiator commemorating his trophies: reminiscent of Edwardian gun-rooms hung with stags' heads, buffalo horns and stuffed tigers.

Up to the present time there has been little money to spare for archaeological pursuits; running water, national health, and welfare services have had to come first. But plans for the archaeological exploration of Nin have been amply set out, together with a comprehensive résumé of what there is and what there was in historic and prehistoric fact.[1] This is a fascinating work which makes a bridge, back through history, to the early Stone Age.

In present-day Nin, the first thing to hold the attention is the 'lower bridge', and the archway of the old city gate. The road by-passes the gateway but pedestrians can walk through. On the right-hand side of the outer wall are primitive carvings symbolic of man's needs — a wine jar, a melon or gourd, and a loaf of bread; together with the face of a man. The one street through the town bifurcates near the top. The forum is at the other end of this straggling lane, and so is the small museum which houses a number of interesting exhibits, though the main treasures, such as the statues of the Roman emperors, are in the museum of Zadar.

These emperors were found in the garden of the Medović house (now the museum) in the eighteenth century; there are four of them. Carved of Carrara marble, they are larger than life, and one has been identified as Augustus, while the other represents Tiberius as augur. The third, represented as *divus*, is dressed in a toga, without a tunic. Although the head is missing it is thought to be Julius Caesar. According to early records there were at least six of these statues. The backs of each were left rough by the carvers: placed in niches, the backs do not show and it is thought they were mass-produced in Italy and ordered *ad hoc*. Perhaps when the emperors gave themselves divine status, the elders of Aenona hastily ordered them for their large temple, 45 metres long and 21½ metres wide, and the largest yet found in Dalmatia. But emperor-worship was not the first religion by any means.

[1] *Problemi Arkeoloških Istraživanja* by Suić, Batović and Belošević, published by the Archaeological Museum of Zadar, 1968. (Serbo-Croat and English text.)

1 The Roman arena at Pula.

2 Lipicana horses and the fiacre on Brioni.

3 Vrbnik, perched on the cliffs of Krk.

4 A Byzantine icon brought by Orthodox monks when they founded their monastery on Palaciol after escaping persecution in the east.

5 The harbour of Mali Lošinj.

6 The brilliant island costume of Susak.

7 The harbour and city of Rab.

8 Stone collars round the olive trees at Lun, island of Pag.

9 Zavratnica Fjord – a small national park especially for marine life.

10 Mending nets at Kukljica, on Ugljan.

11 Sveti Križ (tenth–eleventh century), built at Nin in the form of a Greek cross.

12 Sv. Nikola near Nin; built on a tumulus, with a lookout tower superimposed on its roof.
13 Juraj Dalmatinac's work in Šibenik Cathedral.
14 The statue of Venus-Anzotika with Priapus, Nin.

15 The Radovan doorway of Trogir Cathedral.

16 Kairos – God of the Fleeting Moment, Trogir.

17 Inland on the island of Šolta.

18 Mosaic (second century A.D.) in the Archaeological Museum, Split.

19 Sv. Ivan, Jelsa (Hvar), with the small window through which fishermen used to make confession.

20 The town of Korčula.

21 Flying staircase to the City Gate and tower of medieval Korčula.

22 The port of Komiža, Vis.

23 Hvar harbour with the Venetian arsenal.

24 The small town of Lastovo faces inwards from the sea.

25 The two harbours, 86 metres below the town of Lastovo.

26 Sveta Marija na Grža. Lastovo.

27 The Loggia, and a typical smoke-stack in Lastovo.

28 Lastovo dwellings.

29 Walls of the Roman 'Villa Rustica'; Polače, Mljet.

30 The monastery of Paleokastritsa, Corfu.

31 The twin bays of Paleokastritsa, Corfu.

32 The monks' walk at Paleokastritsa, Corfu.

The statue of Venus-Anzotica points to the Liburnian worship of fertility in earlier times. Even after the Romans' arrival, Liburnians in Nin and elsewhere probably carried on their own cultural and economic development until the first century A.D. According to Dr. Batović's account, it was probably the stabilizing influence of Augustus's rule that finally starved Liburnian independent culture.

In spite of Nin's great importance during early Croatian rule, not much tangible sign of that period is left. There are, however, two little churches. One of them is Sv. Križ, probably of the eleventh century and built in the form of a Greek cross. It is very bare inside with solid rounded apses, and a tower set over the centre. Externally, blind-arch decoration relieves both the church and cupola. It is an endearing building, almost unchanged since the time of the first Croatian kings. Above the door, the lintel is richly decorated with an interlacing pattern, and there is an inscription mentioning the name of Župan Godežav. This lintel is a good deal older than the church. The inscription on a second stone, found during the excavation of the chapel of Sv. Mihovil which once stood in the forum, mentions the Croatian prince Branimir and a building for the Abbot Theodobert. From this information it is assumed that the Benedictine monastery of St. Ambrose was founded during the time of Branimir (879 – circa 892) and that Theodobertus Abbas was its first incumbent. The present parish church, St. Anselm, stands on the site of the much earlier Romanesque-Gothic cathedral and contains some relics from around A.D. 800. A hexagonal font of the same date now in Split, was also found in Nin and provides a fascinating clue to early Croatian Christianity. Nin, at the beginning of the ninth century, was the Croatian centre for political, cultural, and religious activity. Soon after, the Franks overran the area from the west and Frankish missionaries, working from the Patriarchy of Aquileia, began converting the Croats to Christianity. The prince of Nin, Višeslav, was among the first to be baptized, and the baptismal font was named after him. The Latin inscription reads: 'This font — symbolic for Christianity — receives the weak and gives them light. The crimes inherited from the first man are here washed away. You become a Christian who professes the creed of the eternal Trinity for his salvation'. Harmless enough for the Slavs who, from the little evidence that has been left of their earlier pagan religion, do not seem to have been fanatics. They were practical and realistic; they did not believe in fate but honoured their debts when a plea to a god turned out well. So, the 'you become a Christian and IT will give you salvation' slogan must have sounded normal enough in their ears. Yet the Franks, who pushed the new religion, especially during the reign of Charlemagne, were extremely brutal. Porphyrogenitus wrote of torture in Dalmatia. Frankish soldiers are reputed to have fed the Slav babies to their dogs! Throughout history, spreading the gospel has been the excuse for unspeakable cruelty. The descendants of those early Slavs, during the last war, committed atrocities in the name of religion not on the enemy, but on their own Slav brothers-at-arms.

Not very long after the invasion of the Frankish missionaries, one of the most famous of all Croatian theologians was born, Grgur Ninski, bishop of Nin until 928. Charlemagne, with his christianizing zeal, who ended up by being called 'Holy Roman Emperor', died in 814. The Croats had given the emperor's soldiers a tough time; in the same way Bishop Grgur (backed up, at least on the first occasion in 925, by King Tomislav) gave the metropolitan Church and Pope John X a tough time when they tried to curb the breakaway spirit of the Croat Church. Grgur of Nin fought hard for its independence and for the rights of local priests to conduct the Mass in the Slav language. In 928 Grgur's bishopric was abolished; the reasons were partly politics, partly expediency and partly jealousy on the part of the bishops of Salona and Split. Grgur himself faded from the scene and there is no further record of him. But what he had started did not die out; on the contrary it smouldered until, in the mid-thirteenth century, the Pope yielded so far as to allow the old Slavonic language and Glagolitic writing to be used in the island of Krk. In other areas it was used illegally, and became the sword with which Croats could fight for their ancient moral code and their own cultural heritage. Grgur of Nin is still remembered by most Yugoslavs, and also achieved immortality at the hand of Ivan Meštrović, the sculptor. His giant figure stands outside the

Zlatna Vrata of Diocletian's palace in Split, in literally daily touch with the people of the city. The big toe of one of his sandalled feet has been burnished bright because passers by cannot resist touching it!

On the road from Nin to Zadar is another small church, Sv. Nikola, near Zaton. It stands up on a small hillock all by itself, and is accompanied only by six or seven pine trees. Its hillock was a prehistoric tumulus. The ground plan is like a clover leaf — three bulging apses, and one square quarter where the clover stalk would be. In the Middle Ages a hexagonal tower was built above the central dome and used as an observation-post during the Turkish wars.

Leaving Nin means dragging oneself back to the twentieth century; the contrasting bustle of Zadar almost offends one's sensitivities. It is as well to see Nin before Zadar's grander, harsher remains.

Chapter 6 The islands of Brač and Šolta. The mainland town of Split

The islands between Zadar and Split are numerous but comparatively small, whereas the Split islands, like those nearer Rijeka, are bigger but fewer. Directly facing the port of Split are Brač and its smaller sister Šolta.

Brač, with a present-day population of over 14,000, feels unlike any of the other islands. It feels old. And, strangely, much older than its neighbour Šolta, though the two are almost joined at the narrow passage, only about half a kilometre wide, through which the coastal passenger ships pass.

Brač is the third largest of the Adriatic islands, 394½ square kilometres. It also owns the highest mountain, Vidova Gora, which rises to 778 metres. The island has been continuously inhabited since the Neolithic period. There are various suggestions for the origin of its name; some think it came from the Greek 'Bracchia' which had its origin in the Illyrian word 'Brentos' meaning a stag. Numerous traces of the Stone, Bronze and Iron Ages — flint scrapers and drills, small knives, arrowheads, and tools made of stone — have been found. From the Bronze and Iron Ages, fortifications on the hills inland, known as 'gradine', are like strong, stockaded yards; the burial grounds, or 'gomile', also inland, are round constructions sometimes 30 metres in diameter. But it is not easy for the untrained eye to distinguish what is a cattle byre, fashioned out of chunks of unshaped stone, what is a pile of stone (and nothing more), heaped up to clear a small fertile patch, and what is the genuine stone article of prehistoric times. Burial mounds usually look like unwanted pieces of stone, just like the rest! In those early times all the settlements were inland among the hills where it was safer. Early finds were made at Kopačina, Koštilo, Rat — near Ložišća — Škrip, Nerežišća, Hum and elsewhere; none by the sea. Not until Roman times did the inhabitants settle on the shore. So far, there is no evidence that the Greeks actually colonized and built even one 'polis' on Brač, though, in the past, the defence walls at Škrip in the north-western region were considered to have been Greek. Now, however, the experts think that it was not a Greek settlement but that the Illyrian inhabitants, knowing how to make use of Greek techniques, either built the walls themselves or invited Greek builders to construct them.

The Romans, on the other hand, left plenty of traces both inland and on the coast. In the early Middle Ages the inhabitants once more took to the hilly regions. Times were difficult with pirate invasions — Avar/Slav invasions (which finally destroyed Salona close by), struggles for power between rival royal factions, and the early struggles for Adriatic power by Peter Orseolo and other Venetian expanders. And so it was not until Venice had brought a kind of peace to the Adriatic in the fifteenth century that the people of Brač finally settled again near the sea, and lines of communication between Split and the neighbouring island of Hvar six nautical miles away became easier. A record of those early times left by a Brač writer, Dujan Hranković, mentions twelve of the villages which are still in existence today. In 1405 they had a population of 6,000. The main centre was the inland village of Nerežišća in the middle of the island. During the two following centuries this population was cut by the Plague, particularly in 1425 – 7, and at a count in 1599 the number was only 2,000. During these centuries, the people of Brač were cattle-breeders and farmers, with a few fishermen, but in the sixteenth century a new influx occurred. Battles with the invading Turks near Klis, above Split, and at Poljica, above Omiš, brought exiles to the island, and in the seventeenth century the invading Turks in the Makarska littoral and the Neretva Delta forced further refugees on to Brač. These new inhabitants were warriors and their families who, when they arrived on the island, were granted special privileges. Before this time, Brač had two dialects, what is known as 'ča' and 'ca'; but the newcomers brought a third — the 'što' — to Sumartin, a port on the east of Brač. To this day, the dialect varies considerably from village to village — rather like the Channel Islands where the archaic French of one village is hardly understood by the people of the next. From this time Nerežišće, the old capital of the island, diminished.

Unlike most of the islands, Brač has never had a main town which carries the island's name; neither were there any 'nobility' to rule the area under the Venetians — no 'Livre D'or' as in the island of Corfu. In Brač all power and authority were vested in the working landowners. Being cattle-breeders, they did not build elegant towns and were not interested in becoming noblemen, but preferred to live on their estates from which their money derived. Brač's interior still preserves the atmosphere of a husbanded country estate. A few horses still breed on the uplands, and game-shooting is still a popular pastime in this basically rural community. It was not until 1822 – 3, at the Organization of Communities sponsored under Austro-Hungarian rule, that the legal court finally moved from Nerežišće to Supetar which became the official capital in 1827.

From Split harbour, several ferries go each day to Supetar while Jadrolinija's small passenger/cargo ships call variously at all the Brač ports — Sutivan, Supetar, Spliska (for Škrip, one of the four famous stone quarries), Postira, Pučišća, Povlja, Sumartin, Bobovišća Luka, Milna and Bol. On some days you can make the round of most of the island in one ship, calling in for ten minutes at each port, in a sort of lightning tour.

Brač has a lively and varied economy. Brač stone, the main quarries of which are at Pučišća, Selca, Postira, Spliska, and Donji Humac, has been famous since before Diocletian, the Roman Emperor, built his palace at Split of it; the creamy-white, fine-textured slabs are still sent to many parts of the world. Pučišća, a port on the north of the island, is perhaps most nearly linked with the stone. Here there is a healthy stone-carving industry, developed over the centuries; stone mills, lime kilns, and three large quarries. The doorway of the United Nations H.Q. is of Brač stone; so are the White House in Washington[1] and the altar block in the new Liverpool cathedral; stone from Brač was sent to Vimy Ridge for the Canadian War Memorial, and recently to England by President Tito for the statue of Winston Churchill at Westerham, by the sculptor Oscar Nemon, as well as being used by stonemasons in many parts of Yugoslavia, not least in Trogir Cathedral. Another industry is the canning of fish at Postira and Milna (the annual catch is about 500 tons of fish), the Milna fishing fleet being very modern and equipped with echo-sounders. In Venetian times Milna was also famous for wine, much of which was shipped to Venice. The island supports about 10,000 sheep, horses, donkeys, and mules; has an olive oil industry; grows grapes for wine; fruit and almonds. Sour cherries and mulberries are grown to manufacture a soft-fruit drink, and, anachronistically, the medieval centre of Nerežišća contains a chewing-gum factory!

All the ports of Brač are different, and all are scenically attractive; most have hotels and there are lovely beaches, the most famous being Zlatni Rat (Golden Horn). At Supetar there is a modern hotel complex — just away from the main port, near a curving and sheltered beach — protected from the open sea by pinewoods, yet near enough to see the winking lights of Split at night and the approach and departure of ships from the ports. Supetar, too, is the terminal of the road which leads across the island to Bol on its southern side. A bus connects Supetar with Bol and other villages on the way, every day. The entire journey costs 7 new dinars.

There is an organized sea-swimming establishment at Supetar; these are frequent in Dalmatia where water polo is a keenly contested and national sport. The Romans had the same idea at this particular spot, and the remains of their 'piscina' are still there. Because Supetar only became important in the nineteenth century, there is not the wealth of medieval buildings that one comes to expect in Dalmatian settlements. The one really ancient architec-

[1] Oscar Nemon, Yugoslav born, says that American buildings of that period were done by Italians who used stone from the islands without specifying which one; but he recognizes the stone of the White House as being from Brač.

tural example is the pre-Romanesque church of Sv. Luka, on the road leading to Donji Humac. This is eleventh to twelfth century, and contains, carved on one of its walls, a deeply etched drawing of a sailing ship of the same date.

Nerežišća, 10 kilometres inland and the former capital of Brač, is rich in ancient buildings. Near the early Christian church of Sv. Tudor, reconstructed in the twelfth century, there are fragments from the Roman era, and numerous other early and Venetian relics.

Nerežišća, on a hill, is almost half way across the island. It is the junction of the four roads which radiate to different parts of the coast. One of these roads bumps its way close to the summit of Mount Vidova Gora, where there is a mountain hut with a small restaurant and café. Vidova Gora shocks one's senses with its grandeur, it suddenness, and its remoteness. I went up in early June as the sun was going down. The road petered out, the car came to a stop, and a rough track led the rest of the way to the top. We had travelled from the lush, almost tropical vegetation of the sea-level areas; through the agricultural zone of upland pastures, fields and lavender plantations; through the uncultivated maquis-covered slopes, and emerged, through a belt of fir trees, to a veritable alpine scene. The grass was short and wiry, the stones were small and craggy, and the flowers were myriad. Cushiony flat junipers were covered in rampaging honeysuckle; wild thyme jostled saxifrage and campanulas; coltsfoot nodded to groups of small yellow or pink flowers on threadlike stalks, while thistles and daisies wove fantastic shapes, as if Nature had experienced a sudden urge of liberality when viewing this outpost of wild and virgin plateau. I walked to the edge beside the path and looked straight down the steep, sweeping slope to a dark green valley some fifteen hundred feet below, the tree-line clearly visible. It is so steep that the sceen takes on a dream-like quality and reality recedes from the mind. We continued up the steep path to the very top where there is a fantastic view, a sheer drop of nearly 800 metres to the sea. Other islands — Hvar, Korčula, and even distant Vis — are spread out in the blue Adriatic like a relief map. Outside the small mountain restaurant there are simple stone tables. Local food is deliciously cooked: spit-roast lamb, *ražnići*, *pršut*, local cheese and black olives. The many kinds of drink included whisky! From the summer of 1969 the hotel at Supetar is running a sort of guest-retreat in the forest a hundred metres below this summit with bedrooms, bathrooms and a restaurant, so that visitors can transfer from their hotels to the heights of 'Kneževravan' for a few days.

Immediately below Vidova Gora is Zlatni Rat, the famous beach — pale gold and shaped like a horn, with sea on both sides. The easiest way to get to it is by the road from Bol, but there is a path down from Vidova Gora which takes about 2½ hours to walk. Although the Horn looks golden the small stones, close to, are creamy white. There are pine woods as far as the shore, and a few bamboos in the beach near the water. The reflection of light off the white sea-bed makes the water look intensely blue, and it is always calm on one side, according to the wind. In the pine woods is a Roman sarcophagus and the remains of a Roman reservoir.

Brač, with its inherited tough and uncompromising nature, suffered considerably during the last war. In the coastal tourist resorts there is no evidence of this, but I was reminded of it several times during a leisurely progress across the island from Supetar to Bol. British commandos, stationed on Vis, raided Brač on 1st – 4th June 1944. They ran into strong German opposition and incurred heavy losses; unlike their previous landings on nearby Šolta on 18th March in the same year which were much more successful.

Up on the hilly plateau and through Nerežišća is the stony village of Donji Humac. One or two children were playing in the square, and some women were fetching water from the village well. Water is scarce in Brač except at Bol, though a new scheme bringing it from the mainland will revolutionize the position. In the square, which is wide and flat with stone benches under some giant mulberry trees, a mule stood waiting. The mule's master wandered up. He was a jolly middle-aged man who related, with tremendous graphic detail, his good fortune during the war. He had been lined up, as a boy, with fourteen others to be shot after some skirmishing with German troops. A volley of shooting cracked out and he sank down with the other dead of Donji Humac and lay still until it was safe to creep away to the hills.

The bullet designed for him went through his hat, a fact which still affords him great, if sardonic, amusement!

We refused his warm offer of hospitality and turned back along the road to the village of Pražnice, having seen the remains of a Romanesque church and murals of the thirteenth century in the village, the pre-Romanesque chapel of Sv. Ilija, and traces of a Roman mausoleum on a hill. Pražnice looks even stonier than Donji Humac. It is very high, and the view from the slopes towards the mainland over Pučišća is remarkable. The cottages are squat and have stone slabs on their roofs for tiles; these are securely fixed down against the high winds by a very white and thick cement. At one end of the square stands a tiny and ancient chapel with a solid-looking apse nearly as big as itself; it has some curious stone underpinning as if the road had dug away at its foundations. In these inland hill villages life has not yet changed for the younger inhabitants. In coastal towns and island tourist centres there is a marked difference between older women, who seem almost to prefer the original and harder way of life, and the younger women, who wear modern dress, have a knowledge of the world and make full use of modern kitchen equipment. In the interior of Brač life is still the same for everyone. Old women and young girls still scrub their laundry in the village's communal open-air washing-place while they exchange gossip; one family's news is village property; one family's sick child is everyone's care. And many babies are still wrapped in fine linen swaddling clothes to keep their legs straight and immovable, with a sun-burst of embroidered linen behind their heads.

We turned south along the road, which climbed still further, and came to Gornji Humac. These names are derived from *hum* meaning 'hill' and *donje* and *gornje* meaning 'lower' and 'higher'. It was time for the essential revitalizing qualities of Turkish coffee, and so my guide led the way into the home of his wife's cousin. Everyone always has a connection or relative somewhere at hand, who seem delighted to welcome not only the member of the family, however distant, but any visitors as well. The atmosphere on these occasions is completely familial; no stranger is an outsider and usually other villagers arrive and join in.

In Gornji Humac there are fragments of a Gothic triptych, while near by is a chapel, Sv. Juraj, in the village of Straževnik, recorded in 1111. Around Gornji there are also numerous prehistoric grave mounds, and near Donji is the famous Kopačina cave. The entrance is not far from the road, in a mossy dell. Fortunately it is difficult for anyone who does not know about it to get there, and to reach it involves a few leaps and some crawling through bushes. There are no notices and no guard-rails, and it would be easy to fall in. During excavations and explorations carried out with rope and tackle archaeologists found ample proof that Stone Age man had used the deep cave complex and had left many examples of his everyday tools and utensils. The Archaeological Museum in Split is rich in these and other finds from Brač. My guides held my ankles and exclaimed at the appalling risks they thought I was taking as I lay on my stomach and peered over the edge of the gaping entrance-pit. Now only pigeons live down there, using the ledges as perches. I threw a stone over the edge; it took twenty seconds to land, though it had only fallen to one of the subterranean terraces. Seventy years ago, they will tell you locally, a woman taken in adultery was thrown down by her husband — rough and ready justice in a land where there is one law for men and quite another for their women!

The road finally takes a steep plunge down to sea-level and to Bol on Brač's southern coast, facing the island of Hvar across the strip of sea called the Hvarski Kanal.

Bol is sunny and gentle and seems to have been specially favoured by nature with all the necessities of tourism. It is the one place on the island with fresh-water springs. The people, divided by the mountain range from the rest of the islanders, have developed an individual personality, rather as the people of Lun, on the island of Pag, have grown away from the rest.

Slightly to one side of the little fishing port stands the Dominican monastery, set over the sea on a small promontory. On the further side of the port is Zlatni Rat curving out into the sea. Between are several modern tourist hotels of good quality, and the original, commercial, Hotel Luka on the harbour itself. Bol even has a small night club — the Amphora — suitably intimate and shadowy under its fisherman's décor! The town is old; in 827 it was destroyed by the Saracens, and rebuilt. There are many fragments left from the Early Croatian period, while the middle centuries endowed it with many interesting churches and rich houses. Even the post office is housed in a delectable medieval building with an inner courtyard and an outside stone staircase whose beauty almost takes one's breath away.

The Dominican monastery was founded in 1475. The original church was a simple building with Gothic vaulting, and was later enlarged. There are a number of works of art, and the original portico can be seen in the courtyard. At certain times of day visitors can see the museum. In answer to the bell a nun arrives at the door. For people used to Western monasteries and convents, the appearance of a woman in a monk's preserve strikes a strange note, but in this part of Europe it is normal for a few nuns from a nearby convent to housekeep for the men. The Yugoslav way of life obtains even in the monasteries; women do the chores — it keeps them happy and is all in the right order of things. On the day I was there, one appeared at the door from the kitchen, while another, enveloped in a large white apron and with her hands covered in flour, scuttled into the herb garden in search of rosemary or thyme. A good smell came from the monastery kitchen! These nuns looked happy and had the stocky build of good peasant girls.

The Dominican Brother, with many gold teeth, arrived to show me round. His habit, completely white, was of beautiful fine wool, but underneath he wore a pink shirt, though his socks were white and his shoes were black leather sandals. The museum contains Stone Age knives and flints, medieval paintings, manuscripts and incunabula, and a quantity (about 6,000) of old coins. These coins serve as visual Brač history through the centuries and even underline the fact that Russia was in Brač from 1805 – 6, when for a short time Brač became a Russian naval base for their North Adriatic fleet.

From Bol, the coast road continues as far as Murvica before it peters out. There is a small tavern here and, above the village, Zmajeva Špilja, or Dragon's Cave with a chapel in the interior. Now deserted, there was once a monsatery of hermits attached. The walls of the cave are covered with primitive reliefs carved in the rock-face, the origins of which have not yet been traced.

The people of Brač are cheerful, interested and independent. Delighted to see strangers, they are not easily deflected or coerced. Tourism has come to Brač and will undoubtedly expand, but, to my mind, it will be one of the last places to allow tourism to spoil or alter it. The spirit of the island is much too independent.

Šolta

Close to Brač's western side, almost hand-in-hand like a small sister, lies Šolta. Compact and oval, it is 18 kilometres long and varies between 8 and 2 kilometres in width. Not nearly so mountainous as Brač, Šolta's highest point is 237 metres. Though the two islands are so close — less than half a nautical mile separates them at the sea channel through the Splitska Vrata — little Šolta does not seem to be in any way connected with, or subservient to, the bigger island. The mental atmosphere is like that of a small country parish, preserving its own way of life, while looking out of its windows with interest at the world. Tourism has been developed but kept within bounds, and has not been allowed to overlay the small seaports to which it has, so far, come.

Through the ages, Šolta has received various influxes of new blood in her veins from the exiles arriving on her shores. Political exiles were frequently banished from urban areas in Roman times and sent to live on the islands; it was more humane than throwing people into political prisons, executing them, or drowning them with a stone round their necks. If a man — or even woman — was out of favour he 'took the Chiltern Hundreds' of the times, and took himself off to the pleasant life of a beautiful island!

Šolta was first mentioned in the fourth century B.C. when it was called Olyntha and recorded by Pseudoskylax. The Romans called it 'Solenta', and 'Insula Solentii', and it appeared first as 'Šolta' in the fourteenth-century Split statute. After the Avars destroyed Solin early in the seventh century, some of the Solin refugees came to live on Šolta. Again, in the sixteenth century, when Klis, above Solin, fell to the Turks in 1537, Šolta became host to a yet bigger refugee population. But, far from diluting the individual spirit of the people, the refugees must have brought with them those special qualities which make for exiles who are not prepared to lie down under an alien regime: qualities such as energy, restlessness, an awareness of social injustices and unrest and, perhaps most important, an ability to adapt their lives and 'cut their coats' in order to survive. Women, too, were banished at the height of the Roman Empire's influence: women used to intelligent pursuits and conversation; women with wills of their own, and a determination to follow their own inclinations. To take only two examples, there was the Emperor Diocletian's own daughter, of whom he was inordinately fond and whom he banished for her Christian faith; and there was the case of Gallitta, tried in the Court of Justice before the Emperor, and mentioned by Pliny the Younger in a letter to his friend, Cornelianus. Gallitta, wife of a military tribune, brought disgrace on her husband by committing adultery with a centurion. Rome's hierarchical sense was second to none; would she have been brought to public notice, one wonders, had she gone to bed with a senator? However, the centurion was banished. Gallitta's husband did not wish to lose his wife and tried to keep her in his house, but under Roman law both parties must be charged, and both or none convicted. Gallitta was therefore sentenced, forfeited her dowry, and was banished to an island under the *lex Julia de adulteriis*! Just one of thousands of colourful characters to add her blood to island stock!

There are three holiday centres: at Stomorska, Nečujam and Maslinica. Stomorska's Hotel Olint hangs over the sea at the foot of a steep rise; Nečujam's Hotel Marko Marulić — called after one of Šolta's writers (1450 – 1524) who was respected throughout Europe as a humanist and theologian — is in a wide, open bay with pine woods reaching to the shore; Maslinica's is called the Hotel Avlija. Maslinica is, for me, an enchanting never-never corner. The harbour bay is a narrow inlet, like the inner end of Mali Lošinj. Fishing-nets hang out to dry, the tamarisk trees burgeon, and the small hotel, dating from 1708, looks more like a medieval shooting-lodge than a tourist inn. At one back corner there is a high, solid, seventeenth-century tower with a spiral staircase inside, and rooms at the top. It was a lookout point in the days of the pirates. Quiet and peaceful, with a well-stocked bar, the 'C' category hotel has 48 rooms and 110 beds — the sort of place where an extra bed in the bedroom for a child presents no problem.

These tourist centres are spread round the north of the island with Maslinica, the most sheltered, round the corner to the south-west. There is no road along the coast, but only one through the centre of Šolta which branches to the coastal villages, so that the easiest way to reach any of them is by boat. Passenger boats and car ferries run from Split direct to Šolta, and are used as daily transport for islanders whose work is on the mainland. Once on the island, there are always small boats to take people from one port to the next.

I was in Šolta in October. The weather was warm, bright, and clear. The hotels were preparing to close until next spring, and life to go back to its winter pace. Hotel girls, at the end of their summer's employment, were knee-deep in the sea scrubbing out kitchen utensils; bedding was being aired in the hot sun. We chugged from inlet to inlet in a small motor-boat with an ancient engine that had to be primed with methylated spirits before it could be persuaded to function on paraffin. The sea was choppy, the sky clear and blue, and the gulls wheeled overhead as we skirted the rocky coast. Between Nečujam and Maslinica there is a small cove and village with a population of about 50, called Rogać. This is the name for the carob, or St. John the Baptist tree, of which there are some huge specimens on the hill near the harbour. Rogač is the main port for Grohote, the biggest habitation on the island, which is situated in the middle of the high and extremely fertile central plateau. The village of Grohote is extraordinarily spread out and, although there are rows of stone-built cottages, their lines are broadly spaced, while trees and squares of grass border the broad gravel road. Industries include a small plastics factory. The vineyards were full of grapes on the 1st of October, and grapes are loaded into one's arms 'to try'. The parish church stands where a much earlier one stood from the sixth to the fourteenth centuries. The sacristy of the first had a mosaic floor, still extant under an inch or two of fine gravel. Grohote's houses are prosperous, and there is plenty of electric household machinery including, perhaps sadly, coffee-grinders which powder the beans adequately for the ubiquitous Turkish coffee but not quite so well as the old, hand-turned, Turkish mills.

At the other end of the fertile plateau the road reaches Gornje Selo (Higher Village) and then takes a sharp turn to the left and descends to the port of Stomorska.

Gornje Selo is much more compactly built than Grohote. Small stony lanes wind in and out of each other near the brow of a hill. The colour in the clear sunlight is predominantly silvery-white, with handsome thistles and other chalky plants growing in odd corners, and red geraniums near cottage doors. My guide had some relatives in Gornje. The house was a whole collection of ancient stone cottages strung together with stairs and balconies on the outside walls. It was more a '*Zadruga*' than a family: the widowed mother and her brother at the head, various younger relations lower down the scale, and a favourite son — back on leave with his Australian wife and small golden-haired daughter — the prize exhibit. We all went into a raftered room which was very large and long. It was a store-room for the wine; huge barrels, taller than me, lined the back half, painted a bright red. Against a front wall a long trestle table covered in plaid plastic stood in front of a mirror. It was used as a barber's. There were pin-ups pasted on the ceiling and round the mirror — a convenience for those staring upwards while being shaved! In the other part, string after string of small, egg-shaped tomatoes hung from the ceiling, ready for use during the winter. Home-cured *pršut*, sweet home-produced *prošek* and hot bread were rushed out. The Yugoslavs laugh at themselves for the amount of bread they eat, but good bread really means something to them, and they complain bitterly and loudly if it is not up to standard.

The mother, handsome, high-cheeked and short-nosed, gazed at her son and his Australian wife and told me that she had three children, all in Australia. She cried a little, but she would not go out to see them, though they tried to buy her an air ticket. Even the journey to Split, nine miles away, was too far for her!

Life is small in the villages of Šolta, but every detail is important, and news travels fast. One old lady had just collapsed in her home with '*tlak*' (blood-pressure) chatter, chatter! An old man came up the lane to 'look me over': 'But how is it,' he asked, 'that *Gospođa* speaks our language so awkwardly?' But she is English, it was explained. 'English!' he cried incredulously. 'How is it that *Gospođa* speaks our language so well?' Chatter, chatter, chatter! The ability of foreigners to speak their language is taken as an enormous compliment.

In Stomorska the sun was nearly down. In this latitude the difference between summer and winter light and dark is not so marked. I stayed in a private room overlooking the harbour wall. There were embroidered sheets and pillows, a well-equipped kitchen, and a bathroom. In a wide enamel basin on the bathroom floor a live crayfish, the same orange colour as the embroidery of the sheets, waved his arms languidly as I went in, and winked an eye. A gentleman, no doubt! He was still there when I rose at 5 a.m. to catch the boat to Split.

Split

Of all the Dalmatian coastal towns Split is one of the most important, historically, archaeologically, architecturally and commercially. The great Roman Emperor Diocletian (A.D. 245 – 313) ruled from 284 to 305 and built the fantastically magnificent palace for his retirement, and the mausoleum for his death, at Split, near his birthplace of Salona. The British architect, Robert Adam, got inspiration for some of his most magnificent designs, proportions and façades in Diocletian's palace. Today business men with Mediterranean contacts know the importance of Split as a modern international port of call. Ancient Salona can still be traced in a suburb of Split; Diocletian's palace, with excavations producing more of the original every year, still exhibits the blueprint of Adam's elegant proportion; and the Split Archaeological Museum holds some of the most valuable and interesting remains from both the town and the islands.

A day at Solin — present-day name for Salona — to see the ancient Christian remains, the Roman amphitheatre, the theatre, the thermal baths; a day or more at the Archaeological Museum to see the wealth of statues, stone fragments, Greek coins, and finds of the Stone, Bronze, and Iron Ages; a day, or a week, browsing in and around Diocletian's palace and mausoleum, gazing at the sphinx, marvelling at the strength of walls, the beauty of colonnaded heights, the sturdy and simple elegance of underground halls, the one-time glory of the four main gates of gold, silver, copper and iron (and noting the ingenuity of medieval builders who squeezed their dwellings into the nooks and crannies of the palace to form the town of Split), is less than enough. It is better, perhaps, to make repeated visits from one of the nearby islands, taking several bites at this rich, succulent, historical, and outsized cherry. There is an extraordinary atmosphere in the town — a mixture of twentieth-century commerce and bustle, and the almost tangible echoes of history. The spirit of Roman soldiers, the elegance of Roman wit, the confidence of civil rights under the *Lex Romana*, the swish of togas and the quick padding of sandalled feet, or the measured tread of legions crowd in on the imagination. Layer on layer of history make sense of the extraordinary intensity of conversation and discussion which is commonplace in twentieth-century Split. In the late evening its citizens congregate in a small, open rectangle towards the back of the old town. The noise as one approaches is like the chatter of human starlings. The open space is like a forum. Protocol is involved. On one long side students walk together in groups, putting the world to rights. In the centre, older adults meet their contemporaries, discussing the problems of the times. On the further side, elderly people stroll, or sit on steps while they reminisce, compare grandchildren, exclaim at the youth of today and remember their own youthful exploits. Probably it has not changed much since Roman times!

Another extremely evocative annual occurrence is an open-air performance of *Aida* in front of Diocletian's mausoleum, with its built-in Egyptian sphinx to remind us of Rome's North African strength 2,000 years back.

More than 160 years after the death of Diocletian, his palace saw one of the decisive acts in the downfall of the Roman Empire. A Dalmatian general, Marcellinus, tired of the bickering between the Eastern and Western Empires, declared Dalmatia independent. Marcellinus died violently, and his nephew, Julius Nepos, succeeded him. With an army, a fleet, and a burning ambition, he overthrew the Emperor Glicerius in Rome, sending him off to be Bishop of Salona. Julius Nepos had support from both Byzantium and Gaul; events might have taken a more peaceful turn. But he was overthrown after fifteen months by his general, Orestes, who set his own young son, Romulus, on the throne. Not for long, however, as Odoacer soon annexed the 'purple' altogether from the Romans. Julius Nepos still had a stronghold in Dalmatia. Inside the palace he plotted and schemed to sail on Italy, once more to take command and establish law and order. He reckoned, however, without Glicerius. The latter, only four kilometres away in Salona, was also bitterly plotting a return to the throne of Rome. Glicerius and Julius Nepos confronted each other, and Nepos was killed in the palace in 480. Glicerius moved into the imperial quarters intending, at least, to be ruler of Dalmatia, but he, too, was killed within the year. For a short while Ovidius assumed command, but after only one year Odoacer arrived from Rome, overthrew him, and finally wrested all power from the Western Roman Empire.

What turn of chance gave this beautiful area of the Yugoslav coast a grand-stand view of Roman history? Why did Diocletian build his dream-palace on that particular spot, which was near but not in Salona — a port with its own protected harbour? One theory was offered to me by Dr. Dinko Mirošević who, as a G.P., earned the respect and affection of many villagers on the island of Korčula over a span of thirty years. He has made a special study of Diocletian and over innumerable Turkish coffees, he explained that, on medical evidence which he has detected, Diocletian became a prey to acute rheumatism and, knowing that there were sulphur springs close to Salona, decided to build his palace as a home to retire to and, if possible, to recover in. There are few busts or other portraits of Diocletian from which to deduce his facts, but the doctor has made a close study of coins minted at the time, both of Diocletian and other famous faces. Diocletian was a thin, cadaverous man, yet, suddenly in later representations, his neck and face are gross and blown out. Acute rheumatism can bring in its train a heart condition and dropsy which would account for the change. 'Is it possible that there was a sudden change of fashion in the looks of emperors at that time, and that they were required overnight to look rounded and bloated, like Reuben's women?' No; he had thought about it and ascertained that contemporary coins, made in the same workshops by the same group of artists and bearing other faces, were unchanged in style from previous ones. Dr. Mirošević, now in his seventies, worked with rheumatism cases in sulphur baths in his young medical days. Now, retired and 'in pension', as the national Yugoslav retirement scheme is referred to, he can afford the time to study the medicine of almost 2,000 years ago. In his opinion it was rheumatism which set Diocletian's palace as the milieu for one of the final acts of the failing Roman Empire!

Chapter 7 The island of Hvar

Hvar, situated about half way down the coastline of Yugoslavia, constitutes for many visitors the almost perfect concept of a paradise island; 68 kilometres long and only ten across at its widest points, Hvar nestles between the neighbouring islands of Vis, Brač, and Korčula. It enjoys a particularly mild climate which has been known, recognized and gauged since the early Greeks realized its advantages during their frequent trading calls along the length of the Adriatic sea and settled a colony on it almost two and a half thousand years ago.

The population of the island numbers some 12,000 inhabitants, but this goes up sharply during the summer months with the influx of visitors from all parts of Europe. It is a wonderful place for holidaymakers with several small towns, innumerable coves and beaches with sheltered waters, enough historical and archaeological interest to satisfy the busiest of sight-seers, and splendid connections with the mainland by large boat, hydrofoil, and now car ferry from Drvenik on the mainland to Sućuraj and the island's new asphalt road. As a bonus, and still flaunting their felicitous weather, the hoteliers give concessions and even a money-back guarantee to their guests — if there is more than three hours of rain in the day, 50 per cent, and for each day of snow, fog and sub-zero temperatures, a completely free day!

In some ways, Hvar seems almost 'larger than life': colours on land and on sea tend to glow brighter; Venetian buildings seem almost too real to be true — like an expensive opera set; and Hvar people appear to be more alert and energetic, whether they are planning to turn a deserted canyon-village into a holiday centre, making the most of their agriculture and their fortunate climate, or just talking to tourists. Tourism is older in Hvar and more firmly established, after a span of a hundred years, than in any other Yugoslav centre. On the debit side this does, on occasion, produce the professional but disinterested reaction towards individual tourists which one expects in older holiday countries, but does not normally encounter on the islands where enthusiasm for the individual holidaymaker is usually only equalled by a desire to show hospitality. Hvar's tangible 3 millenniums of history follow that of the other islands: Illyrians, Greeks, Romans and Neretva Slavs shaped events; but these, too, were of a more violent nature than in some other areas.

Aided by Dionysius, the ruler of the Greek city at Syracuse in Sicily, a Greek colonizing expedition set out from Paros, one of the Cyclades islands in the centre of the Aegean sea, to settle on Hvar in 385 – 384 B.C. It is a sobering thought, in the giddy light of twentieth-century jet travel, that a ruler in Sicily 400 years before Christ could have had lines of communication good enough to influence the setting out of immigrants from a small sunbaked island near Delos — the reputed birthplace of Apollo — with only ships and runners to bring him news. Under Dionesius's protection, this early Greek expedition made its first bridgehead on the island which they came to call Pharos, or Faros. The easy explanation for the name of Hvar is that it is a corruption of 'faros' meaning a lighthouse or beacon. And indeed, when the small Greek ships made landfall in the sheltered sound (now called Stari Grad), it must have shone out like a beacon — or Φαρος — after so many days on the sometimes stormy Adriatic. There are, however, other theories, as Dr. Niko Duboković-Nadalini, the Hvar-born historian, says. According to one theory, the name was already in use by the Illyrians before the advent of the Greek settlers. The island has also been called, through the centuries, Lisna, Lesna, and Liesna. The last of these names lends credence to the theory that it comes from a corruption of the archaic Slav verb *lijezati*, meaning 'to come to land' or 'to lie up'.

The centre of life on the island during the Illyrian era was most probably at Pitve (ancient Pityeia), situated on the heights above

Jelsa on the north side of the island. A second habitation may well have centred round Dol, near Stari Grad. Undoubtedly it was at Stari Grad that the Greeks made their headquarters in 385 B.C., their name Pharos changing gradually to the Slav *stari grad* or 'old town'. Most probably a second autonomous Greek settlement called Dimos stood where the present-day capital town of Hvar now lies, south and west of the mountain range which divides it from Stari Grad.

There is proof of a constant habitation in the area since earliest Neolithic times, and many examples of Stone, Bronze and Iron Age tools and works of art are in museums: a sixth-century B.C. Corinthian vase found at Blato on the island of Korčula; copper helmets of the sixth and fifth centuries found at Vičja Luka on Brač; and the famous copper armilla from Hvar itself. On Hvar there are traces of these early men in the caves of Grabak (Grapčeva Špilja) and Sveta Nedelja, and there is even some Neolithic building. An early human skull was found in the bay of Skalezije on Cape Pelegrin, and there is a prehistoric site near Mala Lokva.

Perhaps the greatest authority on the early settlements of the islands is the historian Dr. Grgar Novak, himself born on Hvar. He says that the Aegeo-Cretans sailed the Adriatic long before the first millennium. The poet Alkman mentions the Adriatic Henetans. Herodotus gives the information that the Phocaeans were the first Hellenes in the Adriatic, placing the date in the seventh century B.C. There were commercial connections between the Adriatic coast and Greece in the sixth and fifth centuries B.C., and Dr. Novak writes that he is sure that Corinthian products did not arrive in Bosnia overland, but that Corinth or Corcyra (Corfu), or their colony, Epidamnus (later Dyrrhachium, present-day Albanian Durrës) was sending them to Bosnia by way of the Neretva valley. This entailed ships passing up the coast of Dalmatia as far as Hvar in order to bypass the mountainous Pelješac peninsula and enter the sheltered waters of the Neretvanski Kanal, before making their way up the Neretva river valley with its passage through the mountains to the rich ore and silver deposits of Bosnia. It is a fascinating thought that the Greeks, going northwards, used the same Neretva valley which the later Slavs, going south, used in their overflow to the Adriatic coast and islands.

When the Greeks came, they took over the fertile plains and the workable slopes of the island of Hvar, pushing the Illyrians up into the mountains and towards the eastern end. But the Illyrians made a comeback in 235 B.C. when the island became part of the Illyrian state of King Agron. Shortly after, the Illyrian commander, Demetrius of Faros, seized control. In the meantime Agron's widow, Teuta, Queen of Illyria, was besieging Vis, Dürres, and even Corfu, and gaining ground, but her success was short-lived. A new great power appeared, and the island came under the control of the Romans. There are some fascinating remains of these Greek and Illyrian times. They include some enormous stones, shaped and built into dry-walling. These 'Cyclopean' stone walls, as they are called, are similar to those found in Troy and Mycenae. Without the benefit of modern hoists and cranes the effort involved in their erection must have been gigantic. One such Cyclops wall still exists in the cellar of a private house in Stari Grad. It is 11 metres long and may, originally, have formed part of the Greek town's defence walls. There are two other examples on Hvar. One can be seen on the mountain slopes above Jelsa at Tor, dating from either the fourth or the fifth century B.C. Some of the stones are almost 2 metres long. There is a dry-stone wall and a platform of about 40 square metres. The second example is further east at the same altitude on a mountain spur and has a much bigger platform of about 1,500 square metres. Known as Galesnik, with Illyrian foundations and a Roman superstructure, it was still used as a look-out point in medieval times and is accessible by a road, superimposed by the Romans, on the original Illyrian way. As look-out points and forward strongholds, these two emplacements could hardly be bettered as classic military examples of defence, and they can still be visited today. Not so the wall in the cellar. It is a matter of chance whether the owners of the house will open their doors or not, for, though they are conscious of the almost unique antiquity sheltering under their roof, they are also conscious that a man's home is his private freehold!

From 219 B.C., when the Romans arrived, Hvar seems to have become a backwater. Two Romans villas were excavated which had central heating, and an old well at Dolac, close to Hvar town, is undoubtedly Roman. But, by and large, Roman occupation has not left many traces, except for some coins and a few magnificent mosaics. One lovely example of mosaic flooring exists, in five colours, at Stari Grad under the stone paving of a small street called Srednja ulica — Middle Street. It was uncovered, photographed, and is now covered up again.

Hvar already had many layers of civilization when the peaceful Slavs arrived to settle on the island in the seventh century. These were the tribe of the Neretvani who had come from the hinterland of Europe at the gap carved out by the Neretva river. They overflowed to Metković and the amazingly fertile Nerevta delta, and then began to spread to some of the islands. The quiet rule of the Slav Neretvans lasted, with brief intervals, until late in the eleventh century, but from then on, the island of Hvar lived through the vicissitudes of continual change and strife. Croatian kings, Venetians, late Byzantines, Hungaro-Croatians all came and went. Even the Bosnian king Stjepan Tvrtko I (1353 – 91), Duke Hrvoje Vukčić, a fifteenth-century Duke of Split, and the Republic of Dubrovnik had their turn until, finally, from 1420 until 1797 the Venetians were in continuous power. But even during this time some serious rebellions and some determined demonstrations of Hvar's independence of spirit erupted. The most famous occasion was under the leadership of Matija Ivanić, lasting for four years, from 1510 to 1514. Ivanić was a true leader of the common people. The movement grew until an agreement was finally signed granting commoners and noblemen political class equality. There is still an inscription to be seen marking the 'anno Pacis' on the lintel of the theatre. The people of Hvar have always shown a fierce independence of spirit, sometimes against overwhelming odds and sometimes against their common good. In the Middle Ages a horrifying example of island fratricide occurred in an effort to stop all power being vested in 'the few'. A massacre of the nobles by the plebeians was organized in which the instigators took an oath of insurrection on the cross. Legend says that the Christ-figure's nose bled and that the priest holding the cross fell dead. Nevertheless, the plan went bloodily ahead, and the Venetians took retaliatory action, suppressing the uprising with equal brutality.

Of all the medieval 'people's men' on Hvar, the most outstandingly humane was Petar Hektorović who lived from 1487 to 1572. A poet and writer, he has left a house at Stari Grad, and the ruins of a summer palace in the town of Hvar. The sixteenth-century fortified house at Stari Grad is of massive proportions, stretching along the whole side of a square and facing the sea. It was built on low vaults supported on square pillars, and has a garden with cloisters, like a monastery. Petar Hektorović allowed beggars to use one side of the cloisters, but only for periods of three months at a time, to prevent them becoming soft. One of several inscriptions on the walls states that he built the house, which took twenty years to complete, for the enjoyment of himself and his friends. Within the square enclosed by the cloisters there is a fish-pond. The interesting thing is that the pond contains a particular fish called *ciple*. These fish, particularly good to eat, like to live where a river joins the sea in a mixture of fresh and salt water; Hektorović's pond has water from both sea and land, which arrives through a canal. The family of Hektorović was rich, owning vineyards in Hvar, Brač and Vis, and great groves of olives. One part of Petar's home now houses some ancient artifacts — a wooden ploughshare and a very old wooden press. Petar Hektorović also built a church at Stari Grad; it was begun in 1559 on the site of some Roman baths and dedicated to Sveti Rok. But in some ways his most concrete work is the famous epic *Ribanje i ribarsko Prigovaranje*. It is the first Yugoslav realistic poem and contains vivid word pictures of local fishermen together with a strong flavour of nationalism.

Close to the Hektorović house there is a Dominican monastery and the church of St. Peter. The church contains a picture by Tintoretto showing the body of Christ; two of the figures grouped round are said to have the faces of Petar Hektorović (the donor) and his daughter Lucretia. The poet himself was buried in the wall of the monastery.

The oldest church in Stari Grad is Sv. Ivan which was probably built on the site of an earlier temple. Just behind the church and hidden from view, is a sixth- or seventh-century baptistry, discovered in 1957. Certainly the oldest sacred building on the island, and built for baptism by total immersion, the water-container is over a metre in depth and in the form of a double cross. This too, like the Roman mosaics, having been discovered, is now covered over again until money can be found to expose it safely. Close by is the campanile, of splendid proportions, built at the side of the church. The strong base of this bell-tower once formed the seaward doors to the town. Behind the tower in a little square there is a fascinating primitive Roman stone carving standing on its own like a small rectangular pillar: a small deity holding a reversed torch in his hand. There is also a stone relief showing a Roman ship, which has been built into the interior wall of the campanile. But perhaps the most touching reminder of past incidents lies a few metres away from the town. Down a slight incline are four cypress trees; they are all that is left to mark the spot where a convent used to flourish until the Turks came in 1571 (the year of the Battle of Lepanto), destroying the building and the occupants. The town is rich in works of art, and one beautiful triptych by Francesco da Santacroce, of the Virgin with St. Jerome and St. John Baptist, should be specially mentioned.

Stari Grad, the oldest habitation, but fairly new to tourism, has built its holiday hotels on a wooded point away from the town. Not far away Jelsa, which also enjoys a sheltered situation, is equally new to tourism. The town was first recorded in the fourteenth century under the name of Portus de Pitue — the fishing port of Pitve, or Pityeia, dating from Illyrian times. The water of the sound forms a beautiful natural harbour, and the small town arcs round the closed end. The houses near the harbour are built on what was marshy land until it was drained in the nineteenth century, and the name Jelsa means, according to one theory, the alder tree — a corruption of *jalsa*. There are still many alder trees in this extremely lush and fertile area. Round the main square of Sveti Ivan, the buildings are a good deal older; the oldest is probably of the late thirteenth century. The church of Sv. Ivan (early fourteenth century) is very small and has an attractive oval window with massive stone surrounds, low in the wall. Before the harbour land was drained the sea lapped the walls of the church, and the fishermen came to make their confession through this window while remaining in their small boats. On higher ground is the fortified church of St. Sebastian — one of two in the area. This building was first recorded in 1331 and acquired its fortified apse when the Turks were threatening the Adriatic coast in the fifteenth century. At the base of this tower-like structure, which juts out like the prow of a ship, a huge cistern provided water for the besieged inhabitants, enabling them to stay inside when danger threatened.

So close to Jelsa that they are almost twin towns is the small and beautiful Vrboska. The water is exceptionally blue against the pine trees, and a hotel complex enjoys the perfect situation on the wooded points and inlets. The old town lies at the end of an almost land-locked bay. It is quiet and still has the air of a fishing village. The second of the two fortified churches, Sveta Marija, dominates it from the hill. Its fortified cistern looks even more like an aggressive prow and is connected with the main body of the church by an internal passage. The walls are very tall and have outside platforms, used (with boiling oil, lead, etc.) in early times for repelling the enemy. Near by is the church of Sv. Lovra, housing some outstanding pictures, some of which were formerly in the fortified church. They include works by Titian, Veronese, Bassani, and two painted by Medović, the Dubrovnik painter. There is also an interesting modern work by Marko Rašica, dated 1929.

Vrboska remains timeless. A small, low bridge spans the inlet connecting both sides of the harbour. It all shimmers in the hot sun, impervious to the changing world beyond. If hotel guests want to sample life in nearby Jelsa they go, not by land but by boat. For centuries Vrboska has had a small yard for building wooden ships. Matija Ivanić, the fifteenth-century People's Leader, owned a ship which he moored at Vrboska, and it is thought that he was responsible for building the first breakwater in 1465.

Like the name Jelsa, Vrboska has its roots in botany: *Vrba* is the word for willow and points, yet again, to the fact that this area is never short of water. In fact there is sufficient fresh water to sell to the next island down the coast, Korčula.

From Jelsa and Stari Grad a tarmac road with a bus service connects the towns and villages with the rest of the island. Some of the best wines are grown just outside Jelsa, and in good quantity.

The problem is sometimes not how much wine can be produced, but the price of bottles and corks to contain it. In the old days goat-skins were plentiful and practical! In a metropolis like London, where we throw away empty bottles every day, this shortage of containers seems incredible, but in Hvar it is real enough. In fact, when the widowed owner of some rich vineyards near Jelsa came to visit her emancipated daughter in London, the old lady spent half her time visiting all the churches and the other half rescuing empty bottles from the dustbin, determined to fly them all back with her to Jelsa. It was with the greatest difficulty that her daughter and son-in-law dissuaded her, arguing that the cost of freightage would be too high. Being of sensible peasant stock, this made sense, and she sadly replaced the bottles in the dustbin. Common sense will always prevail in a people who have lived close to the realities of life. The richest modern Adriatic peasant may provide his wife with an all-electric kitchen complete with automatic stoves, grills, coffee-grinders, the lot, but when it comes to the best way of preparing fish or tender meat, he will drive his woman up the garden to the primitive smoke-filled shack to light charcoal because it gives a better flavour. First things come first, and food is one of them.

The ride across the mountain before the drop down to Hvar town on the other side is strikingly beautiful. The steep slopes are intensively cultivated. Thick walls of gathered stone striate the mountain-sides in neat patterns, enclosing the tiny fields like a patchwork quilt. In June the colours are brilliant. There are strips of tender green vineyards, shading to weathered copper where the sulphate spray has dried on the leaves; brilliant patches of blue and mauve where cultivated lavender is in full flower — the scent as strong as the colour. Lavender oil, like oil of rosemary and thyme, is produced as part of the island's economy. It is sold in tiny medicine-bottles, is not expensive, and keeps a fresher fragrance than the commercial 'lavender water'.

Even roadmending takes on a sylvan and unlikely quality in this mountain area. The bus will cheerfully back down the steep road for a kilometre to oblige the large yellow steamroller which, though it bears impressive instructions in the Yugoslav language, often wears a proud, rearing Kentish horse on its breast, with the heraldic motto of 'Invicta' beneath. The roadmen, too, seem unreal. Young, golden, and almost god-like, they pop off after 2 p.m. for their rest, and sit under an olive tree in the patchwork of lavender bushes, for all the world like fresh-faced Illyrian shepherd boys!

The upland villages are built of grey stone with their roof slabs cemented down as they are in neighbouring Brač, a reminder of high winds in winter. On *Veliki Petak* (Good Friday) each year, as for hundreds past, the confraternities or guild brothers of the island go in procession from village to village, starting at midnight and stopping before each church for prayers, chants and a great knocking at the church door with the head of the cross they carry.

For many holiday-makers a visit to Hvar means only the town of Hvar, but I have left this undeniably almost perfect medieval and Venetian centre purposely until last. Its emergence as the civic centre of life on the island is comparatively recent.

It radiates beauty and warmth and spreads upwards from its delicate harbour towards the tops of the steep hills behind, like an amphitheatre round the centre of its stage. The warmth is reflected off the hills and the honey-coloured walls of the Venetian buildings, and the panorama is crowned by two picturesque forts standing guard on the hilltops above.

As is often the case, the most impressive way to arrive in the town is by ship, when the whole of medieval Hvar is in view. The Square, the largest in Dalmatia, measures 4,500 square metres; it is paved with slabs of pale stone, and there is an ancient and decorative well in the middle which has been there since 1520 and is still used. At the far end is the cathedral of Sv. Stjepan with its tall, rectangular campanile standing sentinel beside it; while closer, at the entrance to this great open space, are the loggia (1479, on the site of an earlier one of 1289) with its lovely balustrade and stone obelisks, and the famous Venetian Arsenal with its barrel roof and its wide archway. In front and almost enclosed by a low wall is a minute inner harbour for small boats, and this, too, is decorated with small obelisks. The whole scene—white ship, creamy stone buildings, green mountainside, and palm trees on the quay — is reflected in the bright blue waters of the Adriatic which lap gently in this wide bay, sheltered by the small Pakleni islands.

Behind the port are two massive city walls, crenellated and reinforced with towers at their angles; they sprawl down the steep slope towards the east and west sides of the square. They were built about 1278 and strengthened in the fifteenth century. Of the forts above the town, one stands at the apex of the defence walls and was built by the Spanish in the thirteenth century and strengthened by the Venetians in 1551. The second, called Fort Napoleon, at the top of a higher hill, Sv. Nikola, was built by the French during their brief stay between 1806 and 1813. In fact they built two and called them Batterie *de la droite* and *de la gauche*.

The development of the port began in 1455 and it is known locally as Mandrać, reminiscent of the Greek *mandraki*. In Venetian times the whole of the Eastern Adriatic Fleet used to shelter there in winter. The splendid Arsenal, where the war galleys were kept ready to sail against Turks, Genoese, or any pirate enemies of the State, was developed in 1579 and completed in 1611 (there is an inscription over the lintel), but the structure replaced an earlier one of 1331. The main arch of the barrel-vaulted roof has a span of 10 metres, and the galleys could go straight in from the water; at the side is a smaller vaulted hall which now houses a supermarket; the tins and vegetables and cold meats look a trifle incongruous under the ancient wooden roof. It is above the Arsenal that Hvar's famous theatre, the oldest in Yugoslavia, was built in 1612. It is complete and still perfect. Dating from the time of Shakespeare, it has a delicate wrought iron and marble staircase in the foyer, reminiscent of the Crush Bar at Covent Garden, but more ethereal; with red silk curtains in the two tiers of boxes extending round the auditorium. The stage, with its bare boards, has an indefinable air of old greasepaint about it.

Drama attracted many authors in Hvar, even before this theatre was built. Among them was Hanibal Lucić, whose house is in the eastern part of the town. He wrote the first secular play in Croatian, and he lived between about 1485 and 1553. Then there was the writer Mikša Pelegrinović who wrote Carnival plays; he died in 1563. Marin Gazarović, writing in the first half of the seventeenth century, was another; and Martin Benetević wrote comedies until his death in 1607. Frano Blundović, who was born in Hvar, wrote the first seventeenth-century European novel. These men, and Petar Hektorović of Stari Grad, were extremely cultured and civilized, and their work was being produced on this tiny island in the Adriatic between the end of the 1400s and the beginning of the 1600s.

At the level of the second tier of boxes in Hvar's theatre there is a small gallery housing a collection of modern paintings, and, at the entrance to this gallery, a ship's figure-head in the form of a dragon. Again, the great Battle of Lepanto looms up: this dragon sailed into battle at the prow of a ship from Hvar.

The town holds many works of art which are well listed in the local guide books. By far the most talked about is the gigantic painting of 'The Last Supper', earlier thought to be by Rosselli but now attributed to Ingoli. It spans the wall of the refectory in the Franciscan monastery to the east of the town. The 'Last Supper' is famous partly because of its size — the figures are almost life-sized — and partly because a member of the Rothschild family is reputed to have offered enough gold sovereigns completely to cover the picture if the monks would sell it. Happily it still hangs in the refectory, though, to my mind, it is not nearly as deserving of fame as the pictures at Stari Grad and Vrboska.

Wherever you wander in the town, there is something interesting to see, and the delightful part of it is that you have to discover everything for yourself. There is the Lion of St. Mark, with a closed, not open, book, signifying war, on one of the old houses; the florid Gothic façade of Petar Hektorović's summer home; the clock tower, near the loggia, made from an early defence tower. There is the cathedral which was built in the sixteenth and seventeenth centuries on the site of an old twelfth-century Benedictine abbey; and, on a small hill above the town, there is an open-air theatre where they give summer performances in the ruins of a sixteenth-century Orthodox monastery. This monastery had a hostel for the Greek sailors who came to Hvar on Venetian naval or commercial craft. The Franciscan church, too, has many works of art, including a vivid wooden cross by Leonardo Bassano on which you can still see the damage inflicted by the Turks during a raid, and a polyptych over the high altar by Francesco da Santacroce. Among many interesting inscriptions, one dated 1574 records the date of renovation after damage by Uluz Ali in 1571. A second, dated 1612, was put on the grave of Maria Semitecolo, who was the wife of Piero, a Venetian official. It was partly through her efforts that peace was made eventually between noblemen and commoners after the great rebellion. The broad walk leading to the monastery from the town has a whole series of small grottoes, now empty. They were built in 1720 by Marin Capello, commander of the Venetian Adriatic Fleet, and probably were Stations of the Cross. As for the fascinating little Lapidarium and museum, no one would ever find it in the old church of Sv. Marko and its sacristy without being told, though there is a young girl to take one dinar as an entrance fee from the few who discover this unlabelled collection.

But in other ways Hvar has always had its feet on the ground. It had had democratic rule until the fourteenth century. The municipal laws, recorded and codified as early as 1331 remained in force through the Austrian occupation (1787 – 1806) until the Napoleonic French terminated the island's autonomy during their brief stay. Even at this time a document on the monastery wall proves that there was an agreement for the free passage of Hvar ships sailing under the French flag — a flag of great convenience until the British arrived in the troubled waters of that time. Another decree, issued about 2,000 years earlier, can be seen in the Lapidarium of the Dominican monastery at Stari Grad. This is on a Greek stone of probably the second century B.C. and was issued from Paros, in connection with a plea for help from Hvar, or Pharos, as it then was.

Hvar's sense of commerce, too, has remained lively. It was the people of Hvar who exploited the lavender which grows so well on the mountains and who worked up an export market with French perfumers, though the trade is now declining. They also exploited a prolific wild flower of the Compositae family to make an insecticide. This, too, declined with the advent of DDT, but may well revive in the light of present investigations into the harmful effects of some artificial insecticides.

One unique feature of Hvar, and by far the most important for any student of archaeology, can never be taken away from the island. This is the Grapčeva Špilja — the Graback Cave. To get there, until recently, involved a determined effort: a long, climbing hike either on foot or partly on horse or donkey. Now a good asphalt road has been completed to within twenty minutes' walking distance. A stony but easily negotiable path crosses the top of the mountain and drops to the entrance of Grapčeva Špilja on the other side. Near the entrance are some huge lumps of grey rock which have fallen and re-lodged during the centuries. The caves are about 500 metres above sea level, face south, and overlook the small island of Šćedro. In spite of their height and the steepness of the mountain they must have had a broad, accessible approach from the sea before the rock-falls, and one can imagine the processions of neolithic and eneolithic people who came to them for meetings and cult rites, leaving such fascinating traces — traces of an extraordinarily high degree of culture, three millenniums B.C. or even earlier.

The caves were excavated methodically under the supervision of Dr. Grga Novak[1] who has written a detailed and vivid work on the subject; the book is large and has many coloured and monotone illustrations of the finds. There is also a clear résumé in English at the back of the text.

Now the entrance is closed by a small iron gate and it is necessary to have the key, not to mention candles or a torch, as it is absolutely dark inside. There is a small hall, 13·5 × 5 metres, and a large hall 22 × 23 metres, but this extends and rises on its northern side and there is an increase of width to a total area of 31·5 × 32·75 metres. After Dr. Novak's excavation the shafts through the present floor-level were again filled in, but the finds have been tabulated in their various strata and many can be seen in the Split Archaeological Museum. The interesting fact emerged that it was the lowest stratum, and therefore the oldest, which produced the remains of the highest degree of culture. The finds at this level could be subdivided into three categories. First came rough and coarse pottery, unornamented and probably used for preparing food; this pottery contained much calcite and was

[1] Fellow of the Yugoslav Academy of Arts and Sciences.

usually well dried but unbaked. Second came pottery varying from moderate to good and even excellent workmanship; either unpainted, or decorated with only one colour, and made of well-purified clay tempered with minute particles of calcite. This second category had beautiful forms with carefully smoothed exterior and decorations, either incised or pricked, indented or applied on the potter's wheel. Thirdly, there was painted pottery, regularly shaped and made of purified clay and with very beautiful form. Nearly always, the exterior surface was polished, and sometimes the interior as well. Ornamentation was sometimes on both surfaces. There was also a small amount of fine-quality pottery, both unpainted and undecorated. Dr. Novak draws attention to the fact that latticed ornamentation has also been found in the valley of the Indus, in Egypt of the Amratian period — the date placed as about 5800 B.C. by Flinders Petrie — and in Neolithic sites of the fourth millennium in Cyprus and in Syria.

Very few tools were unearthed in this lowest stratum. Only fifty pieces of tools were found, and only seven large pieces, probably imported to Grapčeva Špilja from neighbouring dwelling-caves for the meetings or cult-rites.

The excavations brought to light stone knives, arrows, bone pins, and a stone hammer. At a depth of 2·23 metres they found a human jaw. Another find was a copper armilla (now in the archaeological museum at Split). From the fourth level downwards, there were many charred animal bones. These were all domestic, with the exception of one horn from a hind and a few boar's bones. In the lowest level all the bones were of small tame ruminants, such as sheep and goat. A horse's bones were found in the sixth stratum and there were sea shells at all levels, indicating the presence of humans. Summing up his work Dr. Novak has written: 'The cave of Grabak, which has clarified to a great extent the history of the Adriatic Regions during a period of two millenniums, has also produced problems.'

He deduces that there was a highly cultivated people, during the earliest era, who came, probably from the mainland, to make sacrifices to their forefathers. They brought offerings, the produce of their fields, etc., in vases. In other words, Grapčeva Špilja was purely a place of worship or cult. Probably they lived in other caves or in buildings similar to the round stone structures (*bunje*) which can still be seen in the mountainous areas on Brač, Hvar and other islands and are similar to the Sardinian *Nuraga*. Another interesting point is that the clay with which pottery in Grabak was made is not found in the vicinity of Grapčeva, although a second cave, Pokrivenik, in the north of the island, does contain this particular clay. It seems a fair assumption that the earliest traceable users of the cave, of a highly civilized order and living many centuries B.C., later left the cave unoccupied. Much later, it seems, the cave was again used, probably for habitation, by humans of a much lower culture.

Dr. Novak, like many other historians, is insistent that there is no concrete proof that the legendary heroes were in a particular part of the Adriatic, or any other Mediterranean area. Homer, it is well known, cared little for geographic location; he was primarily concerned with telling a good story. Other writers on the same themes wrote too long after the events had taken place — if, indeed, they ever did — for accuracy. All the same, for me as a romantic, all the threads take on a pattern of probability, even without substantiation: the Neolithic remains at Osor; the evidence in Vis of the cult of the goddess Artemis (as in Corfu); the ancient Amber route connecting the Baltic with the Adriatic via the Sava/Drava/Drina rivers; the Neolithic coloured pottery which was exported from Thessaly to the Adriatic islands. These for me are good enough pointers, particularly when added to Homer's graphic descriptions of winds, storms, sea-currents, and sheltered anchorages, which fit the Adriatic so well. So much seeming 'myth or miracle' proves to have been founded on concrete fact, from the biblical dividing of the Red Sea by a 'strong east wind' onwards. Since the more recent discoveries at Knossos in Crete, we can actually see the labyrinth, a mural of the Bull Dance, the throne of King Minos, his queen's bath and the course of the running, probably perfumed, water in her private apartments, as well as the remains of grandeur and pillared halls which were devastated by earthquake and fire. Every discovery brings the stories of the legendary heroes closer to reality — Theseus, Odysseus, Jason and the others. These men were highly cultured and civilized in their legendary surroundings, handling objects of great beauty and fine workmanship. The Adriatic and Aegean areas HAD these beautiful things at an astonishingly early period, as evidenced in the finds from Grapčeva Špilja.

Chapter 8 The island of Korcula

The approach to Korčula is either by passenger ship along the coast from Split or Dubrovnik — both have airports — or by car, ferry or small motor-boat across the Korčulanski Kanal, a strip of narrow, deep water which separates the mainland (as represented by the Pelješac peninsula) from the eastern tip of the island. This canal, only 1,200 metres wide at its narrowest point, is overlooked by the fortified city of Korčula. Whether one arrives in the bright morning light or among the purple shadows after sunset, the first sight of the little walled city bearing the island's name is an unforgettable experience. Even the most prosaic of visitors find themselves stilled, while for many people this first impression proves indelible. The island, even more than others, seems to hold an irresistible attraction for many visitors who, once they have discovered its charms, come back year after year. They are drawn by its great beauty and the warm friendliness of its people, and they are also drawn, in some indefinable protective way, by its weaknesses. Tourism is making rapid growth, and there are now large modern hotel complexes in its sheltered bays, as well as the original Hotel Korčula, open summer and winter, on the quay.

The medieval town, designed on a circular pattern, juts out into deep blue water. Its rampart walls and corner towers are built of honey-coloured chunks of ancient stone whose reflections in the sea are dazzling in the sunlight, or mysterious after dark. Inside the defensive walls the town rests on a small pyramidal hill, so that the old tiled roofs appear one above the other, like a montage, climbing up to the central cupola of the cathedral. As the boat eases in, palm trees, bougainvillea and the white marble blocks forming the harbour become distinct, the delicate first-floor pillars and Gothic windows of the Korčula Hotel — an Italianate relic of 1910 — forming a backcloth.

Often, among the crowd waiting on the quay, a group of men will detach themselves, form a circle and start singing. Like the Welsh, Dalmatians do not wait for special occasions, but sing spontaneously as the spirit moves them. The arrival and departure of passenger and cargo ships is always of importance to islanders; it is their lifeline, and the comings and goings of visitors and friends is of perpetual interest. This interest is shared in Korčula by business men, housewives and even nuns, who will wave a tablecloth or blanket in farewell from the walls of their convent to one of their 'chaste sisters' — the Yugoslav term is *častna sestra*.

The island of Korčula is forty-eight kilometres long and in most places only eight kilometres wide. A road runs its entire length from Korčula town in the east to the port of Vela Luka at the extreme western end.

The first recorded inhabitants were Greek, though Stone and Bronze Age man had been there earlier and there are caves and graves from this time. When the Greeks first approached the island, probably in the sixth century B.C. they called it Korkyra Melaina, because the maquis and the thickly pine-wooded slopes gave it such a lush and dark appearance. There is some evidence that these colonizers were Cnidians from eastern Greece, helped by the Greek inhabitants of the other Corcyra (modern Corfu) who had earlier founded the first colony in the Adriatic as a port of call on the trade routes to the west. There are still tangible traces of that period of Korčula's civilization: mentions by Roman and Greek historians; an early sixth-century Corinthian vase; and the famous stone tablet found in the village of Lumbarda. It bears a Greek inscription from the fourth century B.C. and tells of a 'stone city' founded near by, actually naming the Greek settlers, which undoubtedly refers to Korčula itself. This inscribed stone is the oldest of all Yugoslav records. There is some evidence, too, that the Phoenicians came to the island to take advantage of the Korčulanis' skill in shipbuilding, a skill which still obtains today.

As with the other islands, Korčula has passed through many hands. From the third century B.C. the Romans spread eastwards until Tiberius finally annexed the whole Dalmatian area in A.D. 9. Korčula became part of the Roman Empire from 35 B.C., and the remains of Roman habitation have been found at Lumbarda, Prižba and Gradina, near Vela Luka. In 493 Korčula came under the control of the Ostrogoths, but in 555 it became part of Byzantium. Though the Slavs entered the Balkans in the sixth and seventh centuries A.D., arriving on Korčula in the ninth century from the Neretva Valley, Byzantine sovereignty lingered on along the coastal fringe until 1102, so that the Slav influx was a slow trickle rather than a sudden invasion, and the next real struggle for power felt by Korčula was that of the Venetians. They first captured Korčula in 1000 (under the Doge Peter Orseolo) but in 1180 it became part (on the old familiar lines) of the Hungaro-Croatian kingdom, and a statute for the town and island was drawn up in 1214. Between then and 1420 it changed hands several times: Zahumlje[1] to Venice; Venice to Genoa in 1298; Ludovik I (1358); Bosnia (1390); and even Dubrovnik (1413 – 17). But from 1420 to 1797 the island, while retaining its autonomy, was under the Venetians. From 1797 to 1805 it was Austrian; from 1805 to 1813 French; from 1813 to 1815 under the British; from 1815 to 1918 Austrian; and from 1918 until the Treaty of Rapallo in 1920 it had Italian military occupation (as it did in the Second World War) before becoming part of the new kingdom of Yugoslavia. In the second war, after the Italian capitulation, the town was bombed by the Germans before being captured by them. It was finally liberated on 14th September 1944.

Many of the older people on the island still speak Italian as well as Croat — a legacy of the years of domination. But they have no love for any of their aggressors. By far the greatest influence,

[1] Zahumlje really means 'people of the hilly country'. Their land was Hum, or Hlum, a principality in the uplands round Mostar, to the south of Bosnia. It changed its name in the fifteenth century to Hercegovina when Stjepan Vukčić, its ruler, was granted the title of Duke of St. Sava in 1448 by Frederick III or the Pope. The German form of the title was Herzog — hence Hercegovina.

however, came from the long Venetian domination. Venetian buildings and palaces still crowd the town and lend an ancient elegance of infinite good taste.

The plan of the old city within its fortified walls resembles the ribs of a ship converging on the central keel. The ribs are narrow streets running east and west, and the keel, stretching north and south, has, at its northern end, the rectangle in front of the cathedral and the Bishop's Palace, while at the south there is the famous flying staircase leading to the market and the second harbour. Built as it is in this form, there is always a breath of fresh air from the prevailing winds to fan the houses. Every narrow street is crowded with old stone houses, all with beautifully carved doorways and windows and the remains of balconies, but many of them are roofless and have been so for hundreds of years. In 1529, after the plague, the houses were burned to disinfect them, destroying all the wood and leaving the stone. It is only now, with the coming of tourism to provide a little money, that some of them are being restored and turned into apartments. Because they love their town, the Korčulanis are taking care to preserve and replace the elegant stonework in its new role. Stone guttering, for instance, costs a great deal more than metal — up to £5 a metre — but it is traditional and beautiful and used whenever possible: stonecutting has been a craft on the island as long as shipbuilding.

There are several ways of entering the old town, all attractive. The Italian stone steps facing the harbour are one way; the hill on the seaward side up to the ramparts is another; but perhaps the most satisfying first entrance is to walk to the landward side skirting the massive corner towers, up the broad flying staircase which once spanned the moat and enter through the fortifications by the Land Gate. The gate and its tower date from 1384; inside there is a later archway, a triumphal arch built by Leonardo Fascolo to celebrate the Venetians' victory over the Turks at the Battle of Lepanto in 1571. Once under the 'triumph' you find a delicious Lilliputian square, bounded by stone walls, with the sixteenth-century City Hall on one side and an ancient house, turned into a fish restaurant, on the other. No wheeled traffic can enter the city to spoil the pleasures of outdoor café tables — the Turkish coffee, the island wine and the advantages of medieval 'Model Town Planning' and the sea breezes it has provided.

Ahead lies the main square and the cathedral of Sv. Marko. On the way there are two stone pillars; one was used as a whipping-post and the other, deeply grooved from top to bottom, took the shaft of a standard, and was a centre for dispensing important news.

On one occasion, however, a piece of news which was vital for the island's survival was given, not from the standard, but the pulpit of the cathedral. On a Sunday morning, several hundred years ago, the bishop was preaching, and because the men of the island were all away fishing, his congregation was composed of women and children and a few elders. Suddenly, from his vantage-point high in the church, he saw through the window the hostile Turkish fleet approaching. He hurried the women out of church and sent them to fetch their husbands' armour and weapons. Then they stood at the ramparts ready to defend their homes. The Turks, thinking they were mistaken in their assessment of an undefended target, turned their ships and sailed away. A procession every year on 15th August commemorates this day when the Korčulani women saved the island.

The cathedral of Sveti Marko was built between the fourteenth and fifteenth centuries by local masters; it has some interesting, and some lovely, features. It is a mixture of Romanesque, Baroque and Gothic architecture. There are paintings by Tintoretto, Bassano and others, a fourteenth-century font, and a ciborium by Marko Andrijić, a local craftsman who built much of the cathedral. The main doorway is of beautiful proportions, with two stone columns carved in rope pattern with the typical Venetian knot. At the heads of these columns there are primitive and sexually interesting stone carvings of Adam and Eve in quadrupedal position who, in turn, support two couchant lions. The cathedral has a particularly lovely rose window, and an elaborate gable with carved gargoyles. Over an inner doorway there is another stone carving representing a winged angel in armour; it looks very much like a winged St. George vanquishing a dragonish Lucifer.

Some of the most attractive churches in the old town are tiny and were originally private chapels. Most are near the cathedral square but one of the most interesting is the church of All Saints —

Svih Svetih — near the ramparts; it is easy to recognize, as the passage linking the church with the sacristy forms a bridge over the street. The church dates from 1301 and has a magnificent collection of fourteenth-century Cretan icons, quite foreign to Korčula which has always been a Catholic community, and probably brought back by devout pirates.[1] There is also a fifteenth-century silver cross and a sixteenth-century *pieta* by Begarelli. In the sacristy is a chest full of the robes worn by the three brotherhoods, or guilds, of Korčula, and some tremendous candles in their holders which are carried in procession round the old town every year in an awe-inspiring ceremony, on 29th July (St. Theodore's feast day) and on Good Friday, accompanied by the bishop. In the procession one recognizes the master craftsman who has been renewing guttering on a roof, the silversmith and the shoemaker, the stonemason and the boat builder, and perhaps the lawyer who was speaking in court the same morning. There is a solemn dignity about it all. Occasionally the heavy, smoking candles in their sconces are passed from hand to hand to shift the weight; and the entire population turns out to follow or watch.

Just behind the cathedral lies the small eleventh-century chapel of Sv. Petar. Facing it, across the narrow street, is a house with a curious square turret where, according to strong legend, Marco Polo was imprisoned for a number of years after the naval battle in 1298 between the Genoese and the Venetians. Marco Polo was taken prisoner by the Genoese and kept under house arrest in Korčula for some years. It seems likely that this earliest of all travel-writers then worked on his famous book in Korčula. Some people maintain that Polo was actually born on the island and not in Venice. There is an entry in the cathedral's baptismal register dated 1520 recording the name of a Polo, and there are still de Polos living in the old house.

Two museums are worth visiting; the first is the Gradski Musej, housed in a former palace in the cathedral square. It is a fascinating muddle of rooms with an enormous variety of objects. The exhibits range from early Greek remains to photographs and firearms of partisans in the last war, from ancient shipbuilding and navigational instruments to some rather terrible furniture of the

[1] Or perhaps Korčula men who took part in the Cretan war (1645–69.)

last century, and, somewhere, some indifferent examples of mass-produced nineteenth-century English china. The other is the ecclesiastical collection in the fine Bishop's Palace, catalogued and arranged by one of the priests, Father Matijaca. There is also a curious and inexplicable statuette of our Mary Stuart sheltering a crowd of kneeling people beneath her open skirts.

But perhaps the most interesting of all the many links with the past is the Arneri Palace and its family.

This palace faces the cathedral, and the original building stretches right down the street as far as the sea — a whole street for one palace! But, like the others, it was struck by plague and the floors and roof timbers were deliberately burnt to cleanse it. The family have never since found the money to repair it though one of the present three brothers lives in a small portion of the original building. What is left of the greater, ruined part is quite magical. Though trees grow out of the walls at rooftop height in the inner courtyard, the remains of carved armorial bearings, delicate stone stairs, and the chamfered supports for absent floor-beams, all pull one back to the departed grandeur of the sixteenth century.

Today the brothers live simply. One of them is the same rolling-voiced man whom Rebecca West calls 'the Cardinal' in her book *Black Lamb and Grey Falcon.* She met him before the war and, though a generation has passed and Yugoslavia has moved from monarchy to a progressive National Socialism, I don't think he, himself, has changed very much. He is still the father figure, known and respected, and I am proud that, when we meet, he treats me as another daughter. His campaign before the war, as it is now, was 'Water, water for Korčula'. Although the hotels and the inhabitants never run short, because water is brought by boat, it would be far better for life and for agricultue if it could be piped from the mainland; then the washing-machines which are making their appearance in the shops would be of more significance, and the beautiful island could become a little richer from agriculture.

Five kilometres from the town of Korčula is the coastal village of Lumbarda, where the earliest Greek stone was found. The road is flat and runs through vineyards which resemble English country fields with stone boundaries. Often the vines are guarded by the thorns of wild asparagus, for it is here that the famous Grk wine is grown. This wine is extremely heavy, and not white but a deep gold in colour. Its name simply means 'Greek', and the very best Grk can be drunk only in Lumbarda and mostly from private homes. Some Grk, however, is now being bottled in Split for export. Lumbarda also produces very good crayfish, as big as lobster and more delicate in flavour. One of the fishermen, Ivan Jurjević 'Knez', has turned to sculpting 'primitives'. His home is full of his carvings, either in wood or in very white stone. His favourite subjects are peasants and fishermen — perhaps a little old woman carrying driftwood on her bent back or a man with a wine barrel — and his work is becoming well known. He is not the only twentieth-century sculptor from Lumbarda. Frano Kršinić, member of the Yugoslav Academy of Arts, was also born there. His home is now open to the public. Also, in Korčula town, a villa is now a picture gallery of the Korčula artist, Vanka's work.

Lumbarda has a beach of fine shelving sand, unlike much of the coastline which is rocky and deep near the shore. A headland, covered in pine trees, protects the beach, and in autumn if the wind is up the colours are extraordinarily vivid; reddish-brown where the sea churns up the sand; aquamarine shading to green and dark blue as the sea deepens; and a backcloth of forest green from the pines. Van Gogh might well have sacrificed his other ear for it!

On the way back from Lumbarda, near the water-polo pool, there is a solarium, a relic of the British garrison which was stationed on the island during the Napoleonic wars. The broad semicircle of this sitting-out place is lined with stone benches, the floor is paved in stone, and the few steps up from the road are guarded on each side by little dwarfish 'needles', like Cleopatra's. On one face is an inscription in classical lettering telling how Colonel Peter Lowen and his men provided this resting-place for their friends, the natives of Korčula. It is still regarded with sentimental affection by the Korčulanis who would never dream of sitting in it themselves: being born to the sun, it is always the shade which they seek!

The British were there two years. When Captain William Hoste, R.N., under the command of Admiral Fremantle, garrisoned the island of Vis in order to harry the French and break the French blockade of British goods to Dalmatia and central Europe, he sent a sub-force commanded by Peter Lowen to Korčula. Korčula was particularly important because the French occupied the Gulf of Kotor (east of Dubrovnik) until 1813, Dalmatia (which was ceded to France by the Treaty of Pressburg in December 1805) and Napoleon's 'Kingdom of Italy'. Thus the island of Korčula constituted the first threat to French ships sailing west towards home. In fact, from earliest times, the narrow, deep waters of the Korčula straits have seen strife and battle; even that irrepressible pirate Uluz Ali had a go in 1571. It is a fascinating thought that the sea-bed must be littered with the spoils of history. Only recently a Roman galley was found by skin-divers. It lay in sand in the Korčulanski Kanal, full of clay pots and amphorae.

To guard the straits, Col. Lowen built a tower like a Martello at the top of the hill above the town. With thick walls, small windows, and tremendously strong, it has an open platform at the top to accommodate heavy guns. The field of fire commanded the entire length of the straits, with a clear view as far as the island of Hvar to the west and Mljet to the east.

From Korčula a bus travels the length of the island to Vela Luka several times a day, stopping at each village. The first part of the journey is made on the road built by the British garrison to facilitate the moving of heavy guns and ammunition up to their tower on the heights. It rises steeply with hairpin bends and, in a few minutes, is in high, mountainous country, the highest point on the island being 568 metres. Olive groves and terraced vineyards are dotted here and there with small stone-roofed huts. The villages are strung out along the road and, where there is no cultivation, the slopes of the mountains are covered with an extraordinarily rich variety of wild flowers and maquis. Because the island is so narrow, whenever there is a fold or cleft in the mountains, sudden vistas of the sea appear: small vignettes of blue sprinkled with rocks and islets, an occasional glimpse of Mljet or distant Lastovo. The first village is Žrnovo; it looks prosperous, and the *domača* or 'house' wines include a strong white which is almost pink. Pupnat is the next village, straggling along two arms of a right-angle. The houses and the stone-roofed huts for the goats are as grey as the stony land around them; there are few trees and the bare 'karst' or limestone rock has an inhospitable appearance. The women, as they come home with their donkeys at six in the evening, look as if life can be hard and the donkeys — a very small breed — are almost smothered beneath the bundles of fodder collected for the goats and pigs. Often an old grandmother will be struggling along with a bundle as big as any.

Soon the bus passes through Čara, a cheerful and pretty village. Visitors are often surprised to be spoken to in Čara in their own language — perhaps Australian, New Zealand, Bronx-American or German. Many Korčula men emigrate for anything between three and twenty years to reap the benefit of richer industrial areas. But whether they go as labourers, housebuilders, engineers, or artists, and even if they have founded new families, the vast majority of them have the overriding Slav urge to return to the land in which they were born.

At the centre of the island the land changes dramatically and becomes very fertile. Smokvica lies in a broad rich valley, like a cradle in the mountains. The wine factory where Pošip is bottled is at Smokvica. It is a good white wine of about the weight of a riesling. The output is high and methods of production are constantly being improved. A good deal is exported as well as distributed to many parts of Yugoslavia. In the director's office there is a delightful photograph of a beaming Marshal Tito, taken at an open-air celebration when he visited the factory to congratulate them on being awarded the Gold Seal for the quality of their wine.

Some of the families in the villages have very high cheekbones, even for Slavs, and wide-set eyes. It is possible that these features are a legacy from the days of Genghis Khan. As well as the secular delights of Smokvica, there is also an interesting church though the key can prove elusive.

Blato, as well as being the last inland village, is also the biggest. It has a reasonable hotel, a medieval parish church containing two pictures by Santacroce and a large and beautiful open-sided loggia, too large for the present size of the community. Not far from the village are three fourteenth-century chapels, Sv. Marija u Polju (1338), Sv. Martin (1346) and Sv. Mihovil (1346). The four-storeyed Renaissance-Baroque citadel with sculptures decorating its well-preserved walls belonged to the Arneri family; its tower now houses a small museum. The people of Blato are different, and have a way of looking suspiciously at strangers; one can sense, even now, that they underwent a particularly hard time during the war. However, once they have overcome their initial suspicion, they are as warm and friendly as any Dalmatian community.

On the road between Smokvica and Blato a steep track runs down to one of the most perfect small bays, often deserted; this is Brna. It is a small piece of paradise worth the effort of getting there. Down another steep gulley where land meets sea there is a bay of mud which has long been known to benefit sufferers from rheumatism. Recently this black mud has been analysed for its medical value and has been confirmed as rich in many curative ingredients.

At the end of its journey the bus runs down the easy slope to Vela Luka. Two long, low spits of land stretch far out into the sea forming a huge natural harbour; fishing is a sizeable industry at this end of the island. At the land apex of the sheltered water, nestling against the town buildings, are the fishing-boat wharves, and when a passenger ship arrives from Split or one of the other islands it has to nudge its way round to discharge its passengers. On one of the long arms of land there is a sardine and tuna factory which produces some of the best tinned fish available. There are two hotels and some *pensions*. Vela Luka's tourist industry is increasing; it has wonderful swimming and is a perfect place for inexpert sailing enthusiasts. From Vela Luka there is a passenger boat connecting with Split, Hvar, and Lastovo.

Clustered round the eastern tip of Korčula are a number of small islands — delightful satellites, some with fresh-water springs, which can be reached by small boat. Closest to the town is Badija, famous for its fifteenth-century monastery. The island is low, with gently sloping pinewoods, and because it lies partly in shallow water, the colours of the sea lapping the rocks shade through zircon colours to deepest sapphire. Boats tie up against the low marble wall in front of the beautiful Franciscan buildings. Before the war there was a school attached to the monastery. There are records of an earlier Benedictine foundation — scoleum sancti Petri — and four eighth- to ninth-century capitals of stone. Though it has been sacked and plundered many times during the last millennium, the monks, with a strange tenacity, have always managed to return. Since the war the building and the grounds have been used as a summer camp for physical fitness enthusiasts, and there is a small restaurant and open-air *kafana*. Until the last war it was a favourite Sunday morning excursion for Korčula people who went across in boats to attend Mass in the monastery church. The cloisters shine with a soft gold in the sunlight and, in spite of the young gymnasts, there is in these cloisters an air of monastic peace and a feeling of continuation, almost of waiting. Although the monks have scattered it would not be surprising if they found their way back yet again. In 1970 the Vatican was deliberating on this very point.

Beyond Badija there are smaller islands, among them Sutvara which is now deserted but still has a splendid stone jetty and two empty cottages from the time when stoneworkers quarried there. Near the jetty are fig trees, wild grapes, and sea urchins by the score, lying among rocks pitted and bleached by the sea like moon craters. Close at hand there is the tiny island of Vrnik with a few houses and the most famous of all stone quarries. Vrnik's smooth white stone was mentioned in records of the fourth century B.C. It has been used in many parts of the world — in Sacré Coeur in Paris and La Scala, Milan, to mention only two. A third island, Majsan, now uninhabited, had an extremely early settlement on its small surface.

Equally accessible by small boat are the villages bordering the sea on both sides of the Korčulanski Kanal. Facing Korčula, the peninsula is dominated by Sveti Ilija, a bare mountain rising to 3,000 ft. In summer it is possible, with a guide, to walk to its summit overnight and watch the dawn — an awe-inspiring experience.

At the foot of Sveti Ilija is the harbour and tourist resort of Orebić which is the terminal for the Korčula ferry. The peninsula is both wild and fertile, producing vegetables and fruit, oranges and lemons and outstanding wines, including the famous and heady Dingać. Just above Orebić, on the steep slope of the mountain, there is a small Franciscan foundation where three Brothers still live. They make a very potent and curative liqueur which one can sample; the walk down the hill afterwards passes in a flash! In the wilder parts of the peninsula, as in the interior of Korčula, jackals still range; looking like underprivileged small foxes, these animals are among the last of their kind in Europe.

Beyond Orebić towards the tip of the peninsula lie the villages of Kučište, Viganj and Lovišće. Viganj, in particular, is of interest. There are actually two villages, one by the shore and the other, separated from the first by orange and lemon groves, set on the hill above. Referred to locally as a gipsy, or 'pharoahs'' village, the actual word Viganj means a kind of bellows used by gipsy tinkers. Here, as in Kučište and Orebić, there are a number of old houses built by retired sea-captains; each has an especially large top window to afford wide-angled viewing of the sea. Even today Dalmatia's retired mariners buy these houses within sight and sound of ships, and a prolonged blasting of a siren in the canal is more likely to signify a greeting than a warning to other shipping; it usually means that a seagoing relative is passing by.

Lovišće is at the head of the long peninsula and, so far, can only be reached by boat, donkey, or on foot. It is a fascinating small community of self-sufficient fishermen who make their own good wine and oil, sell cipal — one of the best fish in the area — and have developed as a side industry the preserving of figs which they treat with sulphur and dry in shallow racks in the sun. In Lovišće there is a small *pension* and restaurant but no shops in the accepted sense.[1] One can buy wine straight from the *podrum* or cellar, vegetables from a barn if the door is open; and when meat arrives by boat, it is put on an open-air table and sold immediately. Fish is not a problem, as every home houses a fisherman.

Across the water on Korčula itself there are the waterside villages Kneže and Račišće, and some tiny hamlets, half hidden in small bays and beautiful inlets. Among them are at least two stone quarries. The stones are cut and shaped on the spot and taken by boat to their destination, whether it is a new pillar for a balustrade,

[1] Though a supermarket is planned.

a carved stone table, or a new step for the market in Korčula town.

Close to the town the fifteenth-century bell-tower of the Dominican monastery of Sveti Nikola stands out like a landmark close to the sea. A thriving religious community, small in winter but swelled in summer when Brothers come for a holiday to Korčula, lives in the monastery, and the mellow sound of the bells which call them to prayer acts as a time schedule for the rest of Korčula, the second time-keeper being the daily arrival of the two passenger boats.

As well as religious processions, the island has its own dances and folk songs. There are several pageants of particular interest. The first is on 23rd April each year at Blato. It takes the form of a stylized conflict to set movements and is accompanied by the noise and beat of bagpipes and drums. Some of the movements are closely related to old rites and dances in many other parts of Europe; for instance, the Decapitation, anciently linked with fertility. After the symbolic mimes women join the men for the final dance. In Blato, the habit of folk 'minstrels' accompanying themselves on the *gusle* — a stringed instrument like a primitive fiddle — has survived.

The second pageant is Moreška in the town of Korčula and is performed every year on 27th July and, nowadays, at other times for the benefit of tourists and cruise passengers. Moreška is becoming well known; its company have travelled abroad, including a visit early in 1966 to England, where they performed in London, Bristol, Leeds and Cambridge.

The origins of Moreška are lost in antiquity, but it has been preserved through at least three hundred years. Some say it comes from the Spanish struggle with the Moors; a few relate it to the five centuries of strife between the Balkans and Turkey, though only the fringe of this long struggle ever touched Korčula. It is not the legend which counts but the fire and vigour with which the spectacle is presented, when the Red and Black Kings, with their followers, meet to do battle.[1] The two sides march on under their banners to some peculiarly strident music. There is a duologue between the Black King and his captive princess, whom he has

[1] Until a few years ago the Black King and his followers blacked their faces.

abducted from the Red King. Then follows a series of soldierly threats between the two kings and a challenge to the death.

The battle-mimes are performed in a large circle, reds on the outside and blacks on the inner. In all, seven set battles are danced, each quicker in tempo than the last; one false step by a performer and he is likely to be badly slashed. At least a dozen of the metal swords are replaced during the performance when they are bent or broken against another. The sparks literally fly off the weapons; the concentration is stark on the faces of the men; the audience, who may start off with smiles on their faces, gradually stiffen with intensity, their hearts beating quicker with the increasing tempo of the music. Finally the Red King and his men vanquish the black Infidels in a last, frenzied action. Korčula is proud of Moreška and teen-age boys are continuously trained to take the place of older men when the pace becomes too hot for the latter.

The third of these ancient customs takes place at Žrnovo in August and is called a *Kumpanija*. This, though some tourists trickle into the village, is not publicized but is essentially a village affair. In red, black, and white costume the players go through a long dialogue interspersed with dancing and drum-beats. The story involves getting permission from the foreign overlords for the local men to hold their ceremonies. It culminates, after some more dancing in which girls join the men, in a live 'sacrifice'. A young bullock (nowadays happily unconscious from a humane killer) is felled at one swoop with a scimitar. This action, requiring tremendous strength, is performed by the village slaughterman who is dressed, symbolically, all in white. After the kill, which must be done at one blow, his clothes are stained with the sacrificial blood, and the steps down from the square to the road run red with blood and water. The spectators, who are breathless at the moment of killing, crowd round the high priest/butcher as if touching medieval wood in a twentieth-century world. If you speak to the people of Žrnovo about this ancient ceremony, they are slightly embarrassed at what is, after all, a barbarous custom and tell you they won't be attending. Yet, on the day, they all crowd into the square under the plane trees, as if to miss this sacrifice might mean bad luck and bring disaster on their village.

Korčula people have a basic character quite different from the people of other islands, and a marked independence of spirit which sits evenly on their shoulders through good and bad times alike. Indeed, some Yugoslavs from other parts of the republic find the Korčulanis uncomfortable to be with, much as they do the people of Dubrovnik. The truth is that both are different — independent, conscious of their long heritage and tradition.

Korčulanis put first things first. They work extremely hard, starting at five in the morning. They are always conscious of the pull of the seasons and the urgent labour of the harvest or the sea. Thus, in the middle of housebuilding they will suddenly disappear in order to gather the grapes or the olives, or help with the launching of a new boat.

Apart from agriculture the main trades through the centuries have been stone-cutting and shipbuilding; now, exporting wine and canning fish are both growing. Shipbuilding, on the other hand, has been a shrinking industry; whereas a century ago there were many shipyards, there are now only a few. Most of them are small family businesses, but there is still one shipyard employing quite a large number of men. Like all businesses with more than five workers in Yugoslavia, it is run on a co-operative, committee basis.

Some of the finest small boats, however, are made at home. The shape of these small craft, shallow-keeled, broad-beamed and very sturdy, which they use for fishing and as runabouts, has not changed much in a thousand years.

I saw one such boat being built on a stone terrace, sheltered from the sun by a huge grapevine; the boat was over six metres long. Small grandchildren were playing near a table, and the rest of the space was taken up by the raw timbers forming the shell. The master boatbuilder, Gino Tasovac, though '*in pension*', is still acknowledged as one of the finest craftsmen on the Dalmatian coast, his son and son-in-law — both workers at the big shipyard — were helping him. They all live, with their wives and children under the one roof. The mother, tall, handsome, and in command at home, came from Zadar — an unusually long journey for a bride of forty years ago.

As soon as I arrived a decanter of Grk wine appeared, and cakes made with almond flour; these, like sugared almonds, being a speciality of the island. This sort of hospitality is standard everywhere, and anywhere in the Adriatic it would give offence not to accept.

The boat on the terrace had the line and co-ordinated curve peculiar to Korčula boats. Each piece of wood had been carefully selected and shaped by hand with an adze. The prow-piece was a dark mahogany colour and the wood had a natural bifurcation running right and left along the gunwales. The tree it was cut from grew in a corner of the garden; it was *Rogač*, the locust tree. To test the strength of the wood each piece is struck on the ground; if it rings, it is true; if it makes a dull sound it is rejected.

Each boat must be registered and have its number painted on. One gets quite skilled in recognizing the place of origin: KO plus a number for Korčula, SP for Split, and DU for Dubrovnik.

Korčula is a delightful island for yachtsmen, and many tie up in the smaller harbour.

In summer, the Jadrolinija ships call daily from both Split and Dubrovnik and there is also the smaller boat from Split to Vela Luka. Car ferries cross from Orebić to Korčula and back every two or three hours, and the island buses, where applicable, meet and even wait for the boats. One bus leaves Vela Luka at 4.30 in the morning and takes the ferry at 5.30 a.m. to drive to Dubrovnik, arriving at 9.0 a.m. The Jadrolinija ship to Dubrovnik leaves Korčula at 1.0 p.m. and arrives soon after five o'clock, staying overnight in the harbour. It returns the following morning at 8.30 a.m. to call at Korčula at noon before going to Split and Rijeka. This overnight stop of ships in Dubrovnik is particularly useful during the Dubrovnik festival. It is sometimes possible to stay the night on board, after attending a play or concert, before returning to Korčula the next morning.

Many people, however, once they have arrived on the island, do not want to leave it even for a one-night excursion. Their idea of bliss is to live as the Korčulanis do, pottering about the bays, exploring the old city, relaxing in the sun and taking a siesta after a very large lunch. Like the rest of Dalmatia, Korčula people prefer fish to meat and they have some very good island dishes. The shell-fish include the huge *lostura*, a bi-cuspid sometimes eighteen inches long. They stew the flesh with oil, parsley and garlic, and it is very good. But perhaps best of all is a feast on the tiny eggs of *ježine*, the sea urchin. To enjoy these to the full you must sit with your feet dabbling in the sea; and you must also have a friend who knows where to find them and who can dive deep enough to get the best. The urchins are cracked like an egg and the inside washed clean with sea water. Then, with a good deal of red wine and fresh bread, you scrape the delectable orange caviar out of the ridges with a finger — for hours at a time — in the paradisal aphrodisian atmosphere. These are the simple things, and the best. These are the invisible strands which bind people, inescapably, to Korčula.

Chapter 9 **Vis**

Vis, one of the outermost of the Yugoslav Adriatic chain, is an island seen darkly, in the distance, from ships on their journey to Split, Hvar and Korčula. It is mountainous, mysterious, and solitary, lying well out in the open sea. Inhabited since the Neolithic era, it is astonishingly rich in its traces of Greek and Roman civilization, though nothing as yet has come to light bridging the end of the Roman era — about the third century A.D. — and the coming of the Slavs in the eighth, to the ninth century. Vis remains one of the unspoilt natural island-havens, partly because of its less accessible geographical position, and partly because of its role as a strategic area and training-ground. In fact it has been a defence point through history, lying well on the way to the Italian Adriatic coast. British troops defended Vis in Napoleonic times, long before they joined with Marshal Tito and the Yugoslav partisans. Since the war, until the beginning of 1969, it was a restricted area, and only now is fully open to visitors.

A daily ship from Split, calling first at Hvar, then at Vela Luka at the western end of Korčula, docks at the town of Vis, on the north side of the island, and Komiža, its second port, on the west.

Even as the ship approaches, past small islets with flashing lighthouses, to enter the long hilly arms of the bay with Vis town at its head, the island seems veiled in mystery. The cliffs of this northern shore rise vertically from the sea, clouds lie motionless across the upper slopes of the mountains, and from a distance it is all without colour, clothed only in a series of monotones. Yet, inside her metaphysical barriers, Vis burgeons with colour —

oranges and lemons, vineyards, and myriads of wild flowers and aromatic shrubs.

For the traveller who wishes to know the island of Vis well, it is a good idea to disembark from the ship at the first port of call, the reason being that Komiža, facing west, has a certain lotus-eating quality about it emanating from the warmth of the afternoon sun and the glow from the walls which continue to radiate heat long after sunset. Vis town, facing north, is brisker and rich in things to see. At present there is one hotel, open all the year, on the quay where the passenger ship docks. It is comfortable and unpretentious, and you have to go out of one door and in through another, some metres along the quay, to its restaurant. From its bedroom windows, all the town lies spread in front, with the site of the ancient Greek town of Issa facing it across the waters of the harbour. There are four crumbling defence towers towards the open sea; they look, for all the world, like English 'follies' or specially built 'ruins' for an eighteenth-century country estate. They do, in fact, bear English names — George (for King George III), Wellington, Bentinck, and Robertson — and were built by British forces when they were defending Vis (as the area headquarters of British operations) against the Napoleonic French between 1811 and 1814. One section of the town is actually known as 'Engleska Luka' — English Port!

Behind the hotel lies the medieval town with beautiful stone houses, as in so many places where population and importance has dwindled, in varying states of repair. Particularly in Vis, one is apt to walk up the narrow, stone-paved streets with head tilted back to admire the elegant moulding of windows and balconies, and wish one had the money to repair each to its former glory. Yet Vis does not rest on a faded gentility; in its small way, it bustles with twentieth-century activity and commerce and an energy often lacking in Mediterranean areas.

Despite the lack of tourists, a museum has been assembled with efficiency and foresight. The ground floor of the building is entirely devoted to the Second World War and demonstrates just how deeply involved Vis was. When the Italians left in the autumn of 1943 after their capitulation to the Allies, Vis became the main base for naval forces attached to the Yugoslav national army of liberation. Pictures of their craft in the museum point a pathetic finger to the meagreness of the vessels and supplies at their disposal. The British Marine Commando Brigade arrived in January 1944 and stayed until October, carrying out jointly, or alone, various attacks including landings on the islands of Šolta and Brač. At the end of 1943, after Korčula was overrun by German forces, the units of the Yugoslav 26th Division were transferred to Vis, together with coastal artillery. Vis became the strongpoint, the only one, from which to fight that part of the war, and by May 1944 the central, and extremely fertile, high plateau had been turned into an airfield. In June the Yugoslav Supreme Staff arrived, and the mountain of Hum, with caves tucked into its steep sides, became the centre of activities. For some time, at this stage of the war, the British mission to occupied Yugoslavia, headed by Sir Fitzroy Maclean, had been in close contact with Marshal Tito and his partisan forces. Brigadier Maclean had already considered the feasibility of directing future operations from Vis and had visited the island with this in mind. Finally the Germans closed in on Tito and his forces at Drvar in Bosnia. They took to the woods and a Dakota — supplied by the British, and crewed by Russians — landed just long enough to pick up Tito and his staff and fly them to safety in Bari, though Tito's marshal's uniform was captured and displayed with pride in Germany. Very soon Tito and Fitzroy Maclean were on their way by the British destroyer *Blackmore* to Vis. There is a vivid account of it in Fitzroy Maclean's book, *Eastern Approaches*. During the five months Tito and his staff were on Vis they used the caves as their headquarters.

Pictures in the museum show Marshal Tito presiding over cave-conferences; examples of arms, some captured from the Italians and some British, give a vivid idea of just how inadequate numbers and defence weapons were. The photographs showing British troops with Yugoslavs bring one a sense of involvement; instinctively one searches for faces one can recognize. In the pictures of Tito, close to him, are the figures of two men. Their faces are dissimilar, the one closed and steady and the other open and artistic, with a wide and enquiring mouth. One is Ranković, a Serb; the other is Djilas, a Montenegrin — Tito's friends and right-hand men in his times of greatest difficulty. Now both have lost their influence, for very different reasons. But the interesting thing is that it is all written in their faces in these wartime photographs.

Upstairs, the museum is a completely different world — one that involves and surprises just as much. Here is a fantastically rich collection of ancient Greek vases and figurines, Roman tear-bottles and statuettes, all in near-perfect condition. Large and beautifully painted urns are so nearly complete that they seem at first to be copies; small Tanagra figures, of astonishing beauty, vivid line, and with not a chip broken off, seem too complete to be original; yet there they lie on open shelves unguarded even by glass. One of the finest of these small figurines is the statuette showing the eternal abstract problem of Love and Soul, or Eros and Psyche. Love is shown as a woman — almost a mother-figure — while Soul is a boy, half grown to manhood. There are also some beautiful carved stone heads and one bronze representing the head of a goddess, delicate and serene. How is it that these ancient examples of art have been preserved in so much finer condition than in other areas? Again, it must be partly Vis's continuous 'separateness' through the ages; her geographical location so far out to sea.

Present-day Vis, an amalgamation of two small settlements, Luka and Kut, spreads round the squared-off end of the long bay which faces north. On the western arm is the site of the Greek cemetery and ancient Issa which was built in terraces. The settlement was a walled town, and traces of this can still be seen. There are also remains of Roman thermal baths, mosaics, and a Roman theatre, which probably superseded a Greek one, though in the sixteenth century a Franciscan monastery and church were built on the site. The semicircular form of the present Franciscan building is due to the fact that they took advantage of the solid auditorium foundations which had been big enough to seat about 3,000 people. Most of the ancient city still lies underground.

On the eastern arm of the bay the older houses date from the sixteenth to seventeenth centuries and include the Renaissance Gariboldi palace with an inscription of 1552, the summer residence of the poet Marin Gazarović of Hvar (early seventeenth century) and the Dojmi-Delupis house where, among other antiquities, are two headless Roman statues. These came from the 60-metre arcade built when the Romans arrived in 45 B.C. and imposed their own style of architecture on the Greek town of Issa, adorning the long arcaded walk with statued niches. Another of these statues, one of the finest in existence and probably representing the Emperor Vespasian, has been in the Kunsthistorisches Museum in Vienna for about a century; perhaps, with the opening up of Vis, it will be returned to its historical setting.

The island was never fortified as a whole, protection being provided by individual fortifications such as towers, and even fortified private houses. One splendid example is the seventeenth-century tower standing at the head of the bay, which belonged to the Perasti family. You can walk right round it and find no entrance, the only way in being from the upper floors! These old families lived in considerable style. The houses were richly furnished and contained many pictures and other works of art, while the women dressed in high fashion, with sumptuous jewellery.

The most notable religious painting is in the church of Our Lady of the Grotto, the parish church, and is *The Madonna With Saints* by Girolamo da Santacroce. Apart from artistic monuments the people of Vis look with great affection on the British cemetery which lies, towards the open sea, on the eastern arm of the harbour-bay. This is a burial-ground of the Napoleonic wars and contains the graves of some Englishmen killed in a battle with the French in 1811. With so much antiquity all around it is almost a shock to come across a thriving plastics factory, working shift hours. Then, a little behind the port, is the wine factory. The island has always produced particularly good wine grapes in quantity, the sandy soil with diluvial deposits being ideally suited to their culture, and the wines made on Vis are among the best in Dalmatia. They include Vugava, a white wine of very good flavour, and a specially good form, from one particular region — of Plavac. The Vis factory produces about 4,000,000 litres of wine each year, but this is not the entire production of the island. The wine grapes arrive in lorries and are tipped straight into concrete open-air pits on the end wall of the factory. Grapes and stalks move into the factory through small and controlled sluice-like entrances straight into de-stalking machines, like large pipes. From there, in a continuous progress, they go through an extractor like a spin-dryer and on to two pressings, the first producing the best wine. The juice is piped upstairs to fermenting vats — huge concrete cisterns —which bubble and boil like witches' cauldrons for three or four days. Home-brewed wine is left two weeks or more! The wine factory lies at the foot of a fertile and sheltered little valley, full of sunshine, and in this valley for the last 100 years they have been producing a special variety of palm tree. The nursery is attached to the wine factory and run under the same management; 40,000 Phoenix Palms (*Phoenix Canariensis*) are produced at one time, and there are also young oranges and lemons, scented mimosa, grafted on to wild stock, and other shrubs, all of which are hardened off in shaded greenhouses from their warm and sunny open-air life, before being sold into less favourable conditions.

The island is about 90 square kilometres in size and lies roughly 45 kilometres from the mainland, but like the others in the chain, it was previously joined to the mainland. Finds of deer bones suggest this and also point to the probability that the land was previously covered in deciduous trees, whereas now there is mostly maquis with a few pines. The earliest human inhabitants were later Stone Age men of between 3000 and 2000 B.C. It is thought that they gave the island the name of Issa, while Bronze and Iron Age man arrived during the second millennium. These were the Indo-European Illyrians who had a working partnership with their compatriots in the Balkans, developing the sea routes between islands and the mainland as far as the Aegean sea for their exchange of trade and culture.

Popular theory suggests that the name Issa comes from a girl in Lesbos whom Apollo loved; it is a pretty thought and the modern Italians, when they occupied the island, followed suit and called it Lissa. But on less romantic lines, Dr. Novak writes that he found evidence that Vis (Issa) was not founded from Lesbos, and points out that the name Issa is very close to the Thessalian Larissa, as well as Antissa in the island of Lesbos.

The discovery of an inscription at the foot of a statue in Vis is important in showing that the cult of the goddess Artemis of

Pherae was practised, and connects the island with countries of the Aegean well before the fourth century B.C. There is, for instance, the temple to the goddess Artemis in Corfu, dating from 585 – 580 B.C.

Theopompus, in the fourth century B.C., said that the Ionian sea was named after Ionius, an Illyrian from Issa who ruled over Vis and its surrounding area. In fact, Issa was minting its own coins in the fourth century — there are plenty of them to be seen in the Split Archaeological Museum — and they bear the effigy of Ionius who was the local hero, pointing to the fact that he reigned before the fourth century B.C. and before Issa was captured by Dionysius the Elder of Syracuse. He settled a Greek colony on Issa in the first decade of the fourth century. The great antique period of the island started at this time, when it was made an autonomous Syracusan colony and continued (after Dionysius) as an independent Greek city state, or *polis*, and even founded colonies of its own. One Greek inscription found on the island tells of a new settlement, founded by the city of Issa, on the island of Korčula. The Illyrians tried to capture the island, but the people of Issa resisted and arranged an alliance with Rome against the Illyrian queen Teuta. Issa was then free to continue its successful trade development, protected by Rome until the confrontation between Caesar and Pompey in 48 – 47 B.C. Issa made the mistake of siding with Pompey and, after Caesar's victory, was reduced from the status of an independent Roman ally to that of a subject of the Roman Empire. Although from this time Issa was overlaid by Roman culture and the Roman way of life, she still managed to remain an outstanding centre of trade in the Adriatic until the final disappearance of Rome's influence. Although no concrete evidence has been found in Vis of just how long this influence lasted, it is known that Byzantium, in the form of individual Byzantine and Roman families, lingered on far later than on the coast of Yugoslavia where the Great Migration of Slav peoples had an earlier effect. Somewhere between the seventh and ninth centuries — historians are not agreed on the date — the Croats came to settle on Vis, which then took on the new name.

In A.D. 997 Vis was attacked by Venetians. When King Ladislas of Naples sold his rights in Dalmatia to Venice in 1409, Vis became a Venetian possession like the island of Hvar, the documents dating from the year 1420. Vis, in fact, became part of the medieval commune of Hvar, remaining under the domination of Venice until 1797, when the Austrians took over until 1805. Then Napoleonic France arrived until, in 1811, Great Britain pushed the French out and, under Captain William Hoste, built a strong naval base from which to pursue British naval policy. When Napoleon was finally defeated Vis, like the rest, went to Austria again.

Besides the forts built by the British in Vis, there are others of Austrian origin; the fort in the middle of the town being the most notable. These forts played their role when Italian naval forces were trounced by the Austrians in 1866 off the north coast of Vis. Not until 1918 did the Italians occupy Vis, and then only for two years. In 1920 it became part of the new kingdom of Yugoslavia. In the Second World War Italy came back, between 1941 and 1943. After their departure in the autumn of that year, Vis became the springboard from which Marshal Tito and the Allies in this sector of the war mounted their operations. The older people of Vis remember, and are ready to discuss for hours, this period which, in many ways, was their proudest moment of history. They were starving, but they were tough; they were lacking in training but they were the only people, at that time, in the whole of Yugoslavia who were free.

The road — a good one — between the town of Vis and the fishing port of Komiža, 18 kilometres away, crosses the huge bowl of the central plateau, high up, and skirts the largest mountain, Hum, before descending suddenly to sea-level down the steep slopes on the outskirts of Komiža. In fact, this journey either by car or by bus is as good a way as any to take a quick look at the interior. It passes tiny hamlets and some very ancient farms with outhouses, rich fields and vineyards, and, above all, deposits one near enough to Titova Špilja to make the approach to the caves — for months Tito's supreme command H.Q. — a matter of minutes rather than hours. There are two caves, set into the living rock of the mountain, Hum, which rises to 587 metres. Now there is a path up to the entrances with levelled stone-paved steps. On either side the steep rocky mountain is covered in maquis and wild flowers. There is a special tall variety of erica with thickly packed pale pink and mauve flowers, misty-blue and dark green rosemary, cistus, juniper and myrtle, with the delicious arbutus berries in September glowing red and orange like jewels as they hang, waiting to be eaten, from their dark green bushes. It is a place of great beauty, though one remembers with awe that at times these berries were all that the partisans had to eat. As one walks up the steps the lizards, which are everywhere, dart away to the sides, in their greens and bronzes, peering back from rocky hideouts between the leaves of small plants, so motionless that they seem to have become pieces of exotic metalwork. During Tito's war-time stay there were no steps, only the rocks and the maquis.

The first cave, to the left of the path, was the conference cave, slightly screened by a dry-stone wall across its mouth. A magnificent strategic view across the mountain shows both the plateau which was turned into the airstrip and the Adriatic. Higher, and to the right, is the cave Tito slept in with, to one side, a small one for his Alsatian dog Tigger — never far away.

Komiža, with quite a big harbour, is tucked into the foot of the mountain, warm, protected, and basking in the apricot rays of the westerly sun. The medieval buildings and fishing cottages hug themselves in the knowledge that their mean temperature in January is 8·9° Centigrade — warmer, they are anxious to tell you, than Hvar and the other famous winter resorts. Their population is about 2½ thousand, almost the same as Vis town; the population of the entire island numbers about 7,000.

The main occupations in Komiža are fishing (and it has a cannery, Fabrika Neptune) and farming, with early vegetables, vineyards, citrus fruit, etc. It is also known for 'St. John's Bread', a type of baby-food made from the beans of the carob tree. There is also a small shipyard where they turn out wooden boats, and a plastics factory. During the war the cannery turned to making medical drugs, and the workshops produced stretchers, sterilizers, etc. As soon as the Italians capitulated in September of 1943 Vis became an evacuation centre for wounded and opened military hospitals. The eastern end of the Bay of Komiža has some wide sand and shingle beaches, and there is a delightful low-built, modern hotel with 100 beds almost at sea-level — the Hotel Biševo.

Komiža came into being when Benedictine monks moved from their home on the small island of Biševo, 5 sea miles away, which was continually attacked by pirates, or *gusari*. They moved to Komiža and built a new home on the hill overlooking the harbour, fortifying the church and monastery with strong walls and two towers, one of which is still standing. In times of danger they could live high up inside the tower. The monks owned the land round the monastery and, being industrious, skilful and kindly, attracted the country people who began to settle near by, forming the port of Komiža. Their church, Sv. Nikola, stands up on the hill, large and imposing; a solid house next to it now houses a small museum. The approach is up a broad, gravel path with hairpin bends. Over the main door of the church is a date 'DCCCL' from the time when the Benedictines built small pre-Romanesque churches and chapels on Vis and the neighbouring islands. The first church of Sv. Nikola, consecrated in 1177 by Pope Alexander III, is only a tiny piece of the main, very large church of today. Locally, Sv. Nikola is referred to as *muster* — dialect word for 'monastery'. In the museum, besides furniture, English plates brought home by sailors in the past and church vestments, there is one small and exquisite picture, a Head of Christ which, they believe, is an original Titian.

Once a year on 6th December, St. Nicholas Day, the people of Komiža bring an old fishing-boat from the beach to the doors of the church and burn it as a thanksgiving for safe journeys at sea. There is a lay Brotherhood in Komiža, whose habit consists of white robes with a red cowl. At present there are twenty-nine Brothers, and on Good Friday they process to the church.

Near the harbour a Venetian tower dating from 1585 was built with money derived from the fishing industry. Built as a fortress, it bears an inscription recording its erection and a coat of arms. Inside, on the ground floor there are some ancient items of fishing gear, and upstairs a picture gallery is devoted to the works ot Đuro Tiljak who painted for many years in Komiža; almost all his works are connected with the sea and fishing. Among other medieval buildings, there is a Renaissance house known as Zanetova Kuća and the Renaissance church of Gospa Gusarica, literally Our Lady of the Pirates, built in the sixteenth century. A picture of the Virgin, it is said, was once stolen by marauding pirates, but they were quite unable to clear the harbour and make off in their craft until they threw the picture overboard; the picture was recovered and returned to the church none the worse for its rough treatment! But the oldest surviving church is a little way out of Komiža; 310 metres above sea level, it is the tenth-century chapel of St Michael.

Of the islands belonging to the commune of Vis, Biševo (5·8 square kilometres) is the nearest, very close to Komiža. It wears an expectant look, like a young girl waiting for her coming-out party, though it has been inhabited for a very long time. The tenth-century church of St. Silvester from the old Benedictine monastery still stands, though reconstructed. There are some small hamlets on Biševo, and upwards of 100 people still live in Mezoporat, mostly wine-growers and fishermen. The fishing grounds are good but, except for the fertile flat centre, most of the island is covered in maquis, with pine woods on the northern side. It is interesting that in 1890 the local peasants became the owners of the land on Biševo, when the former feudal owners were bought out. Its highest peak, Straženica, rises to 240 metres, but its steep shores are the most fascinating part for visitors. There are a number of grottoes and caves, of which two are famous. One is the fantastic Modra Špilja, or Blue Cave. At midday, the rays of the sun penetrate the cave through a hole under the surface of the water, covering everything in a dazzling blue light, while objects under water look startlingly silver, and those on the sea floor show rainbow colours. The second cave is called Medvidina, from the Croat word for the Mediterranean seal. Seal mothers use the cave to shelter their young, and the species is protected by law.

Far out to sea, 25 kilometres west of Vis, is the small island of Svetac, a shortening of the name Sv. Andrija, where there are some Illyrian and Roman traces as well as the remains of Benedictine buildings. The island is wooded with Aleppo pine, and red marble has been quarried there. It is only 4½ square kilometres in area and its north coast is precipitous and up to 200 metres high. There is one small settlement, on the south-east coast, and some wine-growing, fishing and lumbering.

Close to Svetac is the tiny islet of Brusnik, and, further north, 70 kilometres from Komiža, is Jabuka. Both are composed of volcanic rock, and the chaos they temporarily cause to ships' compasses — Jabuka particularly — shows the presence of considerable deposits of iron ore. A compass will swing right across its dial and then swing back. Both islands are uninhabited and very small. Brusnik rises only twelve metres from the sea, and is about 300 metres long, while Jabuka rises steeply and inaccessibly to a height of 96 metres — a lump of almost bare rock sticking out of the sea. The isolation of these islands has developed certain flora and fauna such as *centaurea jabucensis*, and *centaurea chrithmifolia*. There is also a black lizard found on both islands with a double, or forked, tail. I am told that these have also been seen, though rarely, on Vis and Biševo. Because of the special nature of the flora and fauna, as well as their geological and petrological interest, the islands are protected by law. Stone may not be quarried, plants may not be picked or dug up, and the small reptiles may not be caught. Fishing-boats do go there from Vis, but it is as well to know the shores because finding a safe anchorage can be difficult. In calm weather, for those seeking solitude in their own craft, the attraction of uninhabited Adriatic islets may have a pull as strong as the magnetic rock from which they are made.

Vis islanders have strong features — high cheek-bones and often strongly jutting noses — features which suit the men better than the women. Because of their newness to tourism, hotels and holidaymakers' idiosyncrasies are not yet ever-present problems but rather subjects to be discussed with interest and energy. Any subject, in fact, is attacked with vigour, personal involvement and an equality of standpoint; the president of the commune seems just as interested in the individual visitor as the head of tourism. Because of this there is no feeling of 'take it or leave it' as one steps off the boat; immediately one becomes part of the family of Vis. These characteristics must have been invaluable assets during the war when Vis became the 'lone window through which the freedom of their future world could be seen'. Certainly, for many, this wartime role stands out like a beacon. But were the characteristics created out of that necessity of war? I think they have always been in the make-up of Vis's men and women since the Illyrians, the Greeks and the Romans added their love of life and freedom to the matrix of Neolithic man.

Chapter 10 **The island of Lastovo**

From the first moment I set foot on Lastovo I knew it would be impossible for me to write impersonally and objectively about it. It is an island that goes straight to the heart. One becomes involved, bound by its beauty, its gentleness, it simple and transparent friendliness, and by the ups and downs of its life which become one's own direct concern. I know I am not alone in this feeling of involvement. I have met journalists who have spent years travelling and writing cool and unbiased pieces until Lastovo has found the chink in their armour, bowling them over with its simplicity and charm and turning them into hopeful do-gooders in their desire to bring it renewed prosperity.

For Lastovo, like a sleeping princess, has known richer days, and the overriding feeling of Lastovans is regret that so many sons and daughters have felt bound to leave in order to find a more secure living and a materially more rewarding way of life. And yet they are uncertain who has gained. Those who emigrate have their cars, their smarter clothes, their new society and a more luxuriant environment in which to bring up their children, while those who stay toil harder, rise earlier and go more tired to bed. Their amusements are what they can make themselves, plus an oldish film, once or twice a week, in the community hall, a Saturday night 'hop' and, recently, television. But they also know the pleasure of being Dalmatian islanders, and all that goes with this. Fish, fresh caught and grilled on charcoal by their 'woman' — the name for wife is synonymous — wine, trodden and made from personal vineyards; oil obtained from family olive trees and a stone press; and bread baked at home. These pleasures and the joy of entertaining with lavish hospitality — as only the Slavs know how — their own friends in their own homes, plus the knowledge that they stand on equal terms with anyone, these are birthrights not easily given up. This is why, after five or ten or thirty years, the exiles bring back their earnings from their adopted countries, thankfully to settle once again in the homes that have waited for their return. Older, richer, and at last fulfilled, they take part happily once more in the life of their own cultural background, glad to share the harder chores of those they knew as children. But in the modern world of commerce Lastovo *must* find a money-spinner. To be an Isle of the Blest is not enough.

It lies, within distant view of Korčula, well out to sea and usually bathed in a curious misty yet sunny light; withdrawn like a village bride in national costume on the morning of her wedding. As with so many of the Adriatic islands, its history goes back to the late Stone and early Bronze Ages. The Ancient Greeks called it Ladesta, but in the tenth century the Byzantine king Constantine Porphyrogenitus named it Lastobon. Legend says that Odysseus was detained by Calypso in Lastovo for seven years — in such a lovely island, one can scarcely blame him. According to Homer, Ogygia, the mythical island of Calypso, was seventeen days by small sailing boat from Scheria, now generally accepted to be Corfu. Allowing for the prevailing sea currents and adverse winds culminating in the great storm which blew him off his course for Ithaca and into Corfu and the arms of Nausicaa, it would appear to be about right!

Anything seems possible in Lastovo, and in its fairy-tale way even arrival and departure at the present time is like a game of blind man's buff! The only passenger boat arrives from Split, by way of Hvar and Vela Luka, after dark and departs, the same way, before daylight. It is impossible, therefore, to see where they have set you down until the next day's sunrise suddenly produces a scene of ravishing beauty. An ancient bus meets the ship, which arrives at the one sizeable port, Ubli, as does the mini-bus which belongs to the only hotel. The ship disembarks its passengers among a crowd of white-uniformed sailors of the Yugoslav Navy who are searching for friends coming to join them during their months of national service. The Jadrolinija Line passenger ship then douses her lights and prepares to sleep, nudging the small naval craft in the next berth. Civilian passengers are whisked away up grassy, tree-hung lanes and over the hills to the town of Lastovo, 10 kilometres away, or else to the Hotel Solitudo, 3 kilometres from the port. The ride is accomplished in either case in pitch darkness. In the morning one discovers that the hotel, at sea-level, has balconies giving directly on to a dazzling blue channel of sea-water, crossed by an old wooden bridge to the green pine woods of the islet of Prežba on the other side. At the water's edge are three or four fishermen's cottages — and that is all. Their rowing-boats may be drawn up almost to the cottage doors while they, perhaps, mend a net or have their 'marenda'.

The Hotel Solitudo seems to lie at the paradise gate of World's End. Forgotten by the business end of tourism, it quietly plies its trade and is full in July and August. The waters it flanks are known as 'Zlatan Prolaz' — Golden Passage — and connect two beautiful bays. The colours in the brilliant sunshine of this lagoon of blue sea and virgin land outdo picture postcards by miles, and one begins to have an inkling of Lastovo's strength of character in the thought that this is one of the few peaceful spots still left for those who seek the ultimate in quiet and beauty.

The island of Lastovo is the main centre of an archipelago consisting of 46 islands, most of them minute and uninhabited, though the fishing-boats of Lastovo are always popping round them, making use of their rich fishing grounds and sheltered coves. Lastovo itself is mountainous, its highest point reaching 417 metres. Thickly covered in pine woods and maquis, much of the terrain is still in its natural and untouched state, and the flowers are lovely. In October, when I was there, the fragile little autumn-flowering cyclamen and the brilliant yellow autumn crocus were in full bloom in the undergrowth. Between the wild hills are fertile and intensively cultivated valleys with vineyards, olive trees, fruit, gourds and grain. The whole island only covers 52½ square kilometres; it lies 13 kilometres south of Korčula, well out into the Adriatic, and has a gentle climate, the mean temperature for January being 8·3° C. and for August 24·9° C.

Of the two habitations, Lastovo in the north is the main cultural centre, the second being the port of Ubli. The town of Lastovo is built high on the southern slope of a steep hill, facing inland over a broad and fertile valley. There is no hotel, but accommodation in private houses. The sea, though so close, is on the reverse side of this hill and 86 metres below, so that the town is invisible from the sea and from Korčula, across the Lastovski Kanal. In medieval times it was safer for a rich and cultured community to build high up and Lastovo became both very rich and very cultured during the Middle Centuries.

In early times Lastovo was a Byzantine possession; then in the seventh or eighth centuries the Slavs arrived from the Neretva valley. It was first captured by the Venetians in 998 with great ferocity under the Doge Orseolo II. He destroyed the fortified tower on the hilltop overlooking the town, but it was replaced by later Venetians. In the eleventh and twelfth centuries Lastovo formed part of the Duchy of Zahumlje (the dynasty based in Bosnia). The pre-Romanesque chapel of St. Luke near the town dates from that time, or possibly earlier. In 1185 the island came under the Hungarian/Croatian kings, but it went back to Zahumlje in 1221. Finally in 1252 it joined the State of Dubrovnik, and in 1310 got its own statute, obtaining communal autonomy. From this time the town really prospered, and during the following centuries some beautiful houses were built. Some of them are still occupied by the same families. It was during these years, too, that no fewer than twenty-nine churches were built.

The parish church is dedicated to the Saints Cosmos and Damion, and the oldest part dates from the fourteenth century, the Renaissance aisles and the spindle-shaped bell structures being added in 1473. Just underneath the top bell is a small medieval head, said to be that of the Serbian king Uroš II and given to Lastovo in gratitude for their help, though the details are vague. King Uroš II reigned from 1282 to 1321.

The church on Sunday morning was full, though most of the congregation were either elderly or very young. There was a full choir of girl's voices which sang, hidden, from the balcony — lusty, loud singing with a strong Slav flavour to it, and produced from the throat. Just inside the main door of the church there is a beautiful and celebrated bronze holy-water vessel decorated with relief motifs. Dating from the fifteenth century, it was given by Dobrić Dobričević, that same Boninus de Boninis who was the printer from Lastovo responsible for illustrating *Aesop's Fables* in 1486, a copy of which is in the museum at Visovac (see Chapter 4). There is a painting on the high altar of the Saints Cosmos and Damion by Giovanni Lanfranco and also a *pietà* of interest dated 1545. The oldest chapel in the town is the Romanesque Sv. Ivan, 1350, a little above the parish church, but no longer in use. Just below is the enchanting and minute Sv. Marija Na Grža — St. Mary on the Stones! It was constructed by a fifteenth-century priest of Lastovo, Marin Vlahanović. His arms and an inscription can be seen over the door. A service is still held there on the day of Velika Gospa, 15th August. The altar is simple, but there is a very beautiful triptych by an unknown painter of Dubrovnik (which, in the style of the sixteenth century, brings an overlying Byzantine tone to the Renaissance picture). Close by this small church is the loggia, also of the fifteenth century — a rugged open-sided little construction which was used for council meetings in the old days. Perhaps as a gentle reminder to the independent spirit of Lastovo, the republic of Dubrovnik also built a church, dedicating it to Sv. Vlaha (St. Blaise), Dubrovnik's patron saint. In the church there is a baroque painting of the saint holding in his right hand a model of the town of Dubrovnik; painted on wood it is dated 1605. Of the other churches, I very much like the small St. Anthony's (fourteenth century or earlier). Solid and dependable, like the saint, it has a stone doorway with some lovely decoration and, on top of the roof at the eastern end, a small stone figure of St. Anthony who carries a little stone child, all tucked up in his Franciscan habit. The other well-known church is the second St. Mary — Sveta Marija u Polju — St. Mary in the Field. The actual field is now a cemetery, but the church is from an earlier time, about 1450. There are some very grand tombs and, in the manner of the Dalmatians, cheerful-looking photographs on the headstones of some of those who have been interred. The church, built of stone with some fine stone carving, is outstanding for the large altar-painting, the work of Pier Francesco Bissolo, presented by the same printer and engraver of Lastovo, Dobrić Dobričević.

It consists of beautifully grouped figures round the Virgin and Child, who are sitting on a simple marble throne approached by a few steps. Round the steps, on a chequered marble floor, are figures of saints — among them St. Jeremy and St. Sebastian — with, kneeling, a middle-aged man with the face of Dobrić Dobričević and (really the centrepiece) a girl, said to be his daughter, seated at the foot of the steps while she sings and plays the lute. The girl's expression and the draping of her clothes are hauntingly beautiful.

Viewed from the church Lastovo looks a higgledy-piggledy mass of old honey-coloured stone houses. No town-planning here, but only expediency as the guiding factor. There are three hundred and eighty houses, and the population numbers 1,200. The wine factory, which with the school and social centre are among the few modern buildings, lies at the head of the valley below the town. Just outside its doors stretches the sportsground and football pitch. A match engenders tremendous excitement. Both sides wear impeccable shirts and shorts and half the town turns out to watch. These are joyous games, with schoolboys avid to retrieve the ball when it goes out of play. Football is a universal passion in Yugoslavia, but nowhere, perhaps, is it more intense than in the smaller communities.

One of the biggest excitements in Lastovo is the return of sons and daughters from overseas. One evening in a steep cobbled street I was warmly greeted by an old lady, dressed in black and leading a donkey. '*Dobro Došli*,' she said, and kissed me on both cheeks. It turned out that she thought I was Marija, returned from Australia. 'But you are exactly like Marija,' she said, still doubting, so that when we parted I almost thought I was indeed Marija! Round a corner in a second street were some young people, smartly dressed, and a few older ones on the fringe. They were drinking *rakija* and chattering, while their children played near by. It was a party to celebrate the homecoming on holiday of several sons and daughters who had emigrated. In the warm evening the stone-paved streets — all steps and slopes — and the paved areas outside the houses are favourite places for gossiping. Donkeys are the only traffic, and neighbours gather from their several houses to exchange the latest news.

The family closeness of the Lastovo community — practically all of them seem to be related — is almost tangible. Drawn in, one shares in their delights, just as, witnessing a mother's farewell to an emigrating son, the tears sting one's own eyes in sympathy.

Although many of the houses are small by modern standards, they are built of good cut-stone blocks, with carved and shaped lintels and architraves. After the harvest the exteriors are brilliant. Gourds and maize, marrows and pumpkins are put outside to dry in the strong sun. Strings of small, egg-shaped tomatoes (*pomidori*) and black grapes — a special firm variety — hang in long strings from the walls of the houses. When matured in the sun they will keep, firm and juicy, until Christmas. And, of course, there are always the flowers — geraniums, hydrangeas, jasmine, and roses, and many succulent plants. The colours, yellow and orange, pink, red, and bluey-black, glow against the warm stone and the particular blue-green paint which Lastovo seems to favour for its woodwork, making a kaleidoscope as vivid as a child's painting-book.

Some of the houses, however, are large and lofty, stone-built and stone-roofed, with polished wood floors that echo one's footsteps, huge oak dining-rooms and first-floor boudoirs. These were rich homes which still reveal the extremely cultured living conditions which obtained in this isolated island from the fifteenth century. Among documents in the archives between the fourteenth and nineteenth centuries are many family inventories and testaments indicating the possession of much gold and silver in the houses, together with many works of art and valuable books. In one old house, still occupied by the original family, there is some particularly interesting furniture, including a desk and a huge wardrobe which goes up almost to the lofty ceiling and contains four compartments, the doors of which are carved, the two outer ones sloping at an angle from the front. These celebrated pieces are the Fantela-Fernesi family's, and of museum standard. The old houses usually had a stone courtyard with flower-beds, terraces, stone pillars and pergolas, and carved stone wells with armorial bearings. The costume, too, was very rich, particularly that of the women. The dresses, some 200 years old, are still worn on special days. The many skirts and petticoats are made of fine homespun wool in navy blues and reds. They are finely pleated and some are embroidered, a red hem being appliquéd on to white in a small and elegant pattern. The blouses are of white homespun linen with collars standing up at the back, ruched and embroidered. Lastovo women look well in their national costume; they are mostly extremely good-looking, with well-proportioned faces and large, wide-set eyes. Many have fair skin and blue eyes, which strikes a strange note in this southern land.

During Lastovo's association with Dubrovnik, the republic chose a governor each year. He was known as the Count of Lastovo and lived in the Palace of the Count. It is at the top of the hill and overlooks the small town. Because so many documents have been preserved, it is possible to see with what order and practicality life proceeded in the island. There was a communal granary, and the ruins of a communal stone grinding-mill are still in existence on a level with the parish church. There was a hospital, mentioned in the seventeenth century, a fourteenth-century meat depot and a fish market, and a stone column to which miscreants were tied for small misdemeanours.

At the top of the town there is an asphalt road where the bus waits to take people to Ubli, and at the other side of the road is a square terrace with a few trees and stone seats. The far side of this terrace overlooks the minute fishing-port, 86 metres below — the town's connection with the sea. There is a small steep road, overhung with trees, zig-zagging down to the two sides of the narrow inlet which is lined, not with dwelling houses, but with attractive old stone buildings used for boat stores and fishing nets. The fishermen never lived at the port but climbed up, as they still do, to their houses in the town. The quay is completely open to the north and is only protected from the southerly and less dangerous winds. Whenever the *Bura* or the *Maestral* blow the small boats must be hauled up to safety on the shore.

These days there is no central market for fish; when a fisherman comes to the town with his catch he stands in the street, calls out '*Ribe*', and waits for the housewives to run out to see what he has. Among the favourite delicacies is a slim, long fish with a sword-like nose; it looks like a cross between a rapier and a silver-blue eel and it is called *iglica.* Lastovo is also famed for its large crayfish.

The second settlement, Ubli, bears no relation to Lastovo, though in fact it has the remains of an older habitation. It was a Roman settlement, and there are traces of a Roman church and four huge stone sarcophagi close to the harbour. The island was a place of banishment during Roman times. On the island of Prežba, opposite the delicious Hotel Solitudo, are the remains of a Roman country villa. When the Romans died away, Ubli seems to have been forgotten until the Italians re-created it in the twentieth century. The buildings, put up in 1935 in six months, are ugly and shoddy. There is a barracks, still in use, which, like all military establishments, seems to cast a blight on everything. There is one small shop, with little in it, a café — not always open — a small fish-canning factory, now unhappily closed, and an astonishing lack of 'soul' for such an intensely beautiful spot. Even the Roman remains seem to have given up trying to be interesting. There are only 300 permanent residents in Ubli at present; this fact, and the meagreness of social facilities, seems to reflect the saddest period of the island's history.

After the Napoleonic wars, during which time the British had a garrison on Lastovo under the command of Captain Hoste at Vis, the island went to Austro-Hungary (Congress of Vienna) until 1918. But at the end of the First World War, while the rest of Dalmatia and Croatia tasted freedom at last from foreign domination, Lastovo became part of the price paid to Italy under the Treaty of Rapallo. The trouble started in 1915 over a 'treaty of London' which promised the Italians almost all the Dalmatian coast if they would come into the war on the side of the Allies. This was a fantastic promise, undertaken by Lords Oxford and Grey — a kind of panic-inspiration, and quite iniquitous. Under this arrangement they were bargaining to sell out to yet another foreign power (and one which, as we have seen, was always casting covetous glances towards Dalmatia), the very people for whom the struggle for freedom was most precious. Their fellow Slavs further east in Serbia were at that time fighting the most bitter battles in the Allied cause and the Croatians, having borne the Austro-Hungarian yoke for more than a 100 years, were longing to join in on the side of the Allies. Finally the Treaty of London was annulled at Versailles, but too late for Dalmatia; the rot had set in. Italy nagged at the idea of land gains in the Adriatic — a return, in fact, of some of the Venetian areas. So unhappy were the people of Lastovo after the Treaty of Rapallo that most of the younger able-bodied ones left the island. The Italians tried to repopulate it with a colony of eighty southern Italian fishermen, calling it Luigi Razza after an Italian hero. Needless to say, none of this worked out very well in this intensely Slav community. The Italians took small and petty reprisals. A typical example was that of the seven-year-old boy who came to Lastovo hoping to visit his grandparents, still living on in the island. He was made to stand for five hours while the Italian officials at the port deliberated whether he could have an entry permit or not. At the end of the time they sent him back to Dubrovnik without allowing him to see his relatives. The Italians finally left in September 1943, after an occupation of 22 years. Lastovo, of course, had its partisans, and from time to time saw British commandos as well as Tito's men.

Because of its beauty, its seclusion, and its rich fishing-grounds, it is a favourite spot for small-boat visits from other islands. People of Korčula, for instance, go on a three- or four-day trip to Lastovo and its many small uninhabited bays. Often they come back trailing an openwork crate behind the boat containing crayfish they have bought from local fishermen; like this the crustaceans will live for weeks.

The commune of Lastovo includes Sušac, 23 kilometres due west, uninhabited except for a few fishing families and with a lighthouse; the Palagruža group, 56 kilometres south-west, also uninhabited except for four lighthouse families but said to have the richest of all the fishing grounds. The lighthouse on Palagruža is of great navigational importance; its light can be seen for 80 kilometres. There is also a meteorological station. On the main island, Vela Palagruža, there are Stone Age remains. The islands, formed from limestone and a particular type of rock called pelagosite, have given their name to a species of lizard, *Lacerta Sicula Pelagosae*. A second type, *Lacerta Sicula Adriatica*, is found on Mala Palagruža. There are no trees on the islands, but a few endemic plants, including *Centaurea Federici* and *Muscari Speciosum*, have been classified.

Carnival time at Lastovo lasts for about three days. The women are resplendent in their red and dark-blue national costumes, with gold braiding and goffering, and plaited head-dresses, while the men wear decorated shirts and red and black jackets with gold cross-belting, gold-striped trousers, bow ties and ornamented and cockaded bowler hats. The carnival includes a sort of pageant-play which is performed in the streets with an 'ad lib' story. Characters include an admiral, a doctor and his assistant, a vet, a standard-bearer, and a musician who plays a stringed instrument; also a woman supposed to be the mother of the guy — which is always a Turk. The procession moves about Lastovo, pausing to enact different scenes with supporting songs and dances, and culminating with the trial and burning of the Turkish effigy. Songs are then sung while the church bells ring out. The origin of this was a siege of the island by Turkish pirates in the Middle Ages, when the islanders finally overcame their adversaries. Many of the Lastovo songs are sung in the old dialect of the island — the same Croat dialect that was used by some Lastovo writers, particularly in the seventeenth century — famous men like Ivo Kuzmić, Ivan-Luc and Nikola Antica.

As I have said, one cannot stay in Lastovo, dwelling among its community even for a few days, and remain uninvolved and 'outside' it all. The gentle wisdom and individuality of its life, the reasoned manners and customs of its people, the acceptance of folk-lore and of basic, natural truths which have been handed down through the generations, immediately become one's normal way of life. So I was not surprised when, wandering up the hill in the soft evening light on the second day of my visit, I suddenly found myself part of a family I had never seen before. An irresistible smell of fish grilled in the open on charcoal pervaded the air from a terrace on which an old man, with two companions, was sitting. Catching sight of me, he waved me into the garden, making room beside himself on the bench. He was, perhaps, eighty-five years of age, and he sat bolt upright with a black beret on the back of his head. His eyes were blue with that special wisdom in them of the sea and the soil — a wisdom that comes only to a man who has lived close to nature and its struggles all his life. The table was a stout piece of wood set on an upturned barrel. His wife appeared from the kitchen, where hams and wine-skins hung from the ceiling, and poured wine from a carafe. She did not drink herself, but clinked the bottle against my glass as we murmured the normal '*živeli*' toast.

They spoke of their children, and the grandchildren born in Australia; so far, there were only girls. 'But they must listen to me,' the old man said, and his friends settled back smiling.

'If you wish for male children,' he said, and his look indicated clearly that only a fool would not, 'there are simple rules known to many generations in Lastovo. Before taking his bride to bed, a man must sleep alone for ten nights. The room must be shut, with no windows open, and the shutters closed. He must eat meat, fish, and many eggs, but drink no wine, only water. The woman must be *čista* [1] from six days before and eat cheese, milk, and eggs, but no meat. When the couple are together the moon must be exactly in its quarter, and at the first time the woman must reach climax first. It has long been known to my family and, without doubt, male children will result.' It was obvious that this, to him, was a basic law of nature, as natural as sowing in spring and harvesting in autumn. The conversation turned to other subjects while we drank prošek, home-produced and always delicious, particularly when drunk with a cake straight from the oven. The cake, a Lastovo speciality, was baked with fresh green grapes inside. Sex-linking temporarily faded until, a month or two later, I read a report which reminded me of this seemingly unscientific piece of Lastovo folklore. The report dealt with a Danish veterinary surgeon's theories, and was headlined in the *Sunday Times* 'The time you make love may decide the baby's sex'. This method — not so new in Lastovo — relies on a 'pre-determined conception pattern' and 'days of abstinence' too!

There are so many natural riches in the island that, though its economy is badly in need of a boost, half of one wishes to keep it untouched. The small fish factory is closed because its machinery is too old-fashioned to compete in the modern world; yet, in an

[1] Menstrually clean.

area so rich in fishing grounds, it is an obvious potential source of prosperity. Money for new machinery and expertise in modern methods would keep the younger generation on the island and so preserve Lastovo's particular culture. Lastovo's schools are good; young children can attend from the age of four, though in many Yugoslav schools the normal starting age is seven. The upper forms cover a comprehensive range of subjects, taught by young teachers — I spent some time in the Common Room — with a modern approach. I watched an English language class for eleven- to thirteen-year-olds. The complications of our verbs were being taught orally by conversation. The entire lesson was conducted in English, and the accent was good — a tribute to their teacher who was still in her twenties. What is more, when I talked to the children afterwards in the playground, they understood me very well, answering confidently and politely. The waiter at the hotel, too, spoke good English which he had learnt only at school in Lastovo.

Undoubtedly the island needs some form of help in order to keep its head above the waters of the modern world. But the same independent spirit which made them refuse a takeover bid for their hotel, because it entailed accepting 'foreign' staff from the hotel school in Zagreb, will probably save this outpost of exquisite civilization and folklore from being ruined, even in the process of easing its financial worries.

Chapter 11 **The islands of Mljet, Šipan, Lopud, Koločep, Lokrum; The small islands of Supetar, Mrkan, Bobara, Sv. Nikola; with Budva, and Sv. Stefan. The mainland town of Dubrovnik.**

As the chain of islands trails south-eastwards towards Albania they become smaller in size and in stature. The off-shore islands near Dubrovnik cannot be compared historically with the mighty strongholds near Rijeka, with the isolated cultures of Vis or Lastovo, or the past grandeur of Korčula. But, between the extremes, is the never-never-land quality of the island of Mljet. Long and slim, it lies parallel with the Pelješac Peninsula, and separated by the five sea miles of the Mljetski Kanal; 36 kilometres long and about 5 kilometres wide, its silhouette, seen from ship or mainland, rises and falls in a series of ridges, the highest being 514 metres. It always looks empty of human habitation, but tucked away in sheltered bays and inland valleys there are about a dozen villages with around 2,000 inhabitants between them, the biggest village being Babino Polje, high up in the interior. The western end of Mljet is well within sight of the island of Korčula, 10 sea miles away, while its eastern extremity is only 13 sea miles from Dubrovnik.

Stepping ashore at one of its small ports, one begins to understand why people rave about Mljet, whether they be foreign tourists or Yugoslavs. The normal passenger boat from Dubrovnik proceeds, easing up Mljet's northern coastline, to call at several harbours. But there are also excursion boats from Korčula which land one at the fishing inlet of Pomine at the western tip. Here the path that serves as a quay also acts as a fish reserve, being hollow and filled with ever-changing sea water. When the planks of the pathway are lifted, lobsters and fish blink in the sudden light; it is these fish that the small cottage-restaurant near the shore will serve on request. They sometimes call Mljet the 'Green Island', and indeed it is the greenest of all. The first thing one learns is that there are three inland sea-water lakes leading one to another. From Pomine it is a three-kilometre walk through the woods to the smallest and most inland of these lakes. There is also an asphalt road across the island, and a bus. But to walk through the still green woods is the perfect prologue for one's first breathtaking view of the lakes. In September there are cyclamen in full flower on the banks, and autumn crocus, brilliant, clean, and bright yellow. The heat of the sun draws the scent from the pine trees and sends it wafting about one's head. The insects buzz lazily across the peaty path, and the fact of the bus seems like heresy in a dreamy, lotus-eating land.

On Mljet, one is alone in a small green world. The colours are green-green and blue-blue. It all resembles an arboretum, carefully tended, and carefully arranged to look natural — which in fact it is! Oleanders bend huge flowers over the lakes; fish, intent on their own business, ignore one. It is so devoid of crowds that it is easy to bathe without costume, and without being a 'nudist'.

I arrived, without luggage, intending to stay one day, and it seemed the most natural thing in the world to stay two nights and a day, buying only a toothbrush from the village post-office-cum-store, which has the delicious smell about it of 'all sorts' not encountered since childhood.

On the largest of the lakes is a small island with an ancient Benedictine monastery of some size. In this beautiful building, which has not housed Brothers for some 150 years, they have made the only hotel. One dines in front or under the cloisters or in the vaulted hall, once the wine cellars; one sleeps in the cells — happily quite spacious ones — previously inhabited by the monks. The church attached to the monastery is huge and has a service once a year on 15th August for Velika Gospa. The beautiful creamy stone walls and the elegant broad stairs and terrace seem to have been tailor-made to fit the available space, yet even on this smallest of islets the shady paths and miniature coves, sprinkled with small devotion points from the past, are uncluttered and remote. For those who seek roaring gaiety, Mljet might seem like a purgatory — no bands, no dance floors, no packaged entertainment — but for dreamers and escapists it is near to paradise.

Besides the one hotel, private houses arrange accommodation, some of them at the lake's edge, and, of course, this is the way to take part in the real life of the island. Someone will always be taking his small open boat out to fish, and it is easy to go along too. When the grapes are ready in September, extra pairs of hands are welcome to cut the swags of grapes off the vines and tip them into the open barrels which stand at the edge of the vineyard. Near the lakes, in a broad flat bowl surrounded by wooded slopes, is one of the best areas for wine. Men, women, small children and donkeys all work at harvest time, some packing the fruit into goatskin sacks — still thought by some to be the best. These sacks are fifty to sixty years old, and are handed down in families. The donkeys stand by to carry them, slung on their saddle-bows, to the wine cottage. In this particular vineyard the setting is almost Elysian; round the edge are broad, low 'balconies' or terraces of short flat grass beneath ancient olive trees that bend their twisted bark and cast a lacy shade. Here the donkeys wait, and the gathered fruit. One is welcome to sample as many bunches as one can manage, and it is interesting to listen to the experts evaluating each type. In the evening, when the grapes are in the wine sheds, it is still possible to see a few who prefer to tread the juice out in the age-old way; the action is more gentle than machinery. Every family makes enough wine for itself to last the year; the surplus goes to the co-operative factory. Besides fish, wine and olive oil, Mljet's industry includes growing medicinal herbs.

A large part of the island has been made into a National Park, covering some 3,000 hectares; flora and fauna are protected and the rich, wooded hills carefully preserved so that the wild beauty of the island cannot be destroyed.

Mljet has cultural and historical associations which go deeply into the past. Traces of settlements, graves and an earthwork from the pre-Greek Illyrians have been found close to the biggest lake. The Romans called the island Melita, and its inhabitants gave the Roman Empire a good deal of trouble by continually attacking Roman shipping. The islanders had a splendid vantage-point for this, as the Romans made their main sea-trade route from Epidamnus through the straits between Pelješac and Mljet to Osor, Istria and Italy. Finally things got so bad for the Roman trade route that the Emperor Augustus sent an expedition to quell the nuisance and annexed Mljet. There are solid traces of these Roman times in many parts of the island, but the most astonishing, because of their size, are the remains of a palace, built on the seashore at the small port of Polače. The port is a wide, sheltered bay, facing north and protected by a ring of islets; there are freshwater springs in the village where about sixty people live. The limestone hills behind are wooded or covered in maquis, and a little above the palace walls in the wild scrub are the half-hidden remains of an early Christian basilica.

According to legend, and some old Dubrovnikan writers,[1] the palace belonged to a Sicilian called Agesilaus from Anazarbus. Mljet was a place of banishment for political and religious exiles. It is said that Agesilaus was banished to Mljet by the Emperor Septimius Severus and that later he and his son Oppianus were pardoned by the emperor Caracella after — so the story goes — the emperor had read the son's great idyll *Aleutica*. The palace was built in the style of the grand Roman rustic residences of the Imperial period, with two great towers projecting from the main façade which faced the sea. The cottages of the present-day village have been built, through the ages, in and about the huge Roman building, and the stone from one half of the palace has completely gone into their making. But there is still a great deal left. The outside walls are gigantic, a metre or so thick and about twenty metres high. There are the joist-holes of at least two upper storeys and some beautifully curved and arched windows. There was a huge central hall or nave, a sea portal at the centre of the front façade, and the towers — one at each extremity — which were rounded. Thanks to the skill of those early builders the flat, smallish stone slabs used to form the curve are still in position. Originally, the building must have been enormous.

Mljet's middle history follows closely that of Lastovo. About the year 536 it became part of Byzantium. Later, the Neretva Slavs arrived, followed by the Zahumlje people. Sv. Andrija and Sv.

[1] Razzi and Orbini.

Mihovil, near the main inland village, Babino Polje, date from this time. In 1151 a Zahumlje ruler, Desa, presented the entire island to some Benedictine monks who came from Mount Gargano in Apulia; it was they who built the beautiful monastery and church on the biggest lake.

It is extraordinary how, though it is now an hotel, the monastery has kept its identity and its atmosphere. The entrance to the church is up some broad stone steps to one side of the landing-stage and dining-terrace. The entrance to the hotel bedrooms is also up these steps, and on through an archway to the inner cloisters, so that guests are always passing the beautiful doorway to the church, just as the monks must have done in their day. This church porch, with high reliefs, was built at the end of the twelfth century; the big square portico, a Renaissance structure with a splendid coat of arms, was added later. The sacristy, defence walls, and tower, complete with firing loopholes above the doorway, date from the same time (fifteenth to sixteenth centuries). Mljet was under attack during this period from the famous Turk Uluz-Ali.

In 1333, the *Ban* or ruler of Bosnia, Stjepan Kotromanić (1322 – 53), having acquired the principality of Hum and the lower part of the Neretva valley, presented Mljet to the republic of Dubrovnik. It is an interesting thought that Kotromanić was still at this time a Bogomil by religion although in 1340 he was persuaded by the king of Hungary to become a Catholic. From 1333 Mljet was administered by a count who lived in the Palace of the Count at Babino Polje. In 1345 the island obtained its statute and remained under the protection of Dubrovnik — like Lastovo but unlike its near sister-island Korčula — until 1808 when Napoleon decreed that the republic had ceased to exist. Napoleon reconstituted the whole of the Dalmatian coast and its islands in what he called the Illyrian Provinces, taking the name of the people whose country it had been some 2,500 years previously, and, incidentally, for the first time grouped Serbs, Croats, and Slovenes together more than 100 years before the South Slavs were finally to unite! His audacity was short-lived; the Congress of Vienna and the treaties of 1814 – 15 assigned Dalmatia and Dubrovnik to Austria.

The only way to visit the southern shores of the island is by small motor-boat. Small groups of hotel guests are collected at the landing-stage by one of the islanders who has a boat. The route lies through the narrow entrance to the outer lake, which is very shallow except for a narrow channel past the rocky entrance to the open sea. At this entrance there is an old stone cross, on the seaward side, with an inscription in the Cyrillic script, probably dating from the second half of the thirteenth century. The sea is deep and blue and open as far as Italy on this side of the island. The slopes of Mljet rise steeply, thickly covered in trees, mostly of pine. Again, the island seems almost deserted. About half way along this coastline there is a grotto or 'blue cave' with the same mysterious marine light of Biševo and Capri; to get inside you must choose a calm day and transfer to a small, flat-bottomed boat. From this southern coast it is just possible to see the main village, Babino Polje, up among the hills, but there is no real bay until one reaches the eastern tip of the island. Here there are a couple of extraordinary sounds. One is protected by a barrier reef only inches below the surface, and the other, approached through a deep-water channel, has a landing-stage. Both bays are deserted and edged with the finest white sand like tropical lagoons; pieces of wood, bleached and thrown up by the sea, lie about like tree-skeletons, and wild flowers grow in profusion on the shore and in the woods beyond. These bays have never been changed by man and must still resemble the deserted shores of the far-off days of the Heroes.

And what of St. Paul who was cast up, after a great storm at sea, on Melita as recorded in the Acts of the Apostles (Chapter 27)? There is circumstantial evidence to show that Mljet, then called Melita, was the island in question. The sea, in this part of the Adriatic, can turn from a millpond to a wind-lashed cauldron in a few minutes if the barometer drops. Conditions change with lightning speed, and at least one British frigate has lost her sun-awnings before the ratings could get them down. St. Paul was shipwrecked on the island in a storm such as this, after many days of sailing without sun or stars for navigation. He wrote of being bitten by a snake while piling wood on a great fire, built by the kindly 'barbarians' after the shipwreck and the rain. Mljet, until this century, always had many snakes, including the original viper which moves in vertical curves, like the serpents depicted in early religious paintings. On the other hand, Malta, the rival claimant for St. Paul's visit, has never seen a snake. Again, St. Paul spoke of the kindness of the inhabitants under their leader, Publius, but calls them barbarians; at this time, Mljet's inhabitants were attacking the lawful purposes of Rome, and any people opposing the empire were referred to as 'barbarians'. Furthermore, St. Paul wrote in verse 27 'we were driven up and down in Adria', and he describes in verses 41 – 4, the creek where two seas met, and how they drove the ship in and aground while the 'hinder part was broken with the violence of the waves'. Now, the channel from the open sea to the first of the lakes (or inland seas) was, until a deeper channel was cut in 1960, only 0·60 metres deep; no wonder the ship carrying St. Paul stuck fast in front and was broken at the stern! There was a bridge across the channel until 1960, and the ebb and flow of the current was used in the Middle Ages to power mills.

Through the centuries, Mljet has kept a rich culture alive, partly because of the monastic settlement which attracted writers such as Mavro Vetranović-Čavčić (1482 – 1576), Basilij Gradić, of the same period, Mavro Orbini who lived until 1614, Anzelmo Banduri (1671 – 1743) and Ignat Đurđević (1675 – 1737). The island's costume was also particularly rich and colourful, though I have yet to see it worn on the island; there are examples in the Ethnographical Museum in Belgrade. The snakes which were so prevalent on the island now no longer exist. Instead, there are quantities of mongoose which were imported before the First World War to rid the island of reptiles. There are no longer snakes for the mongoose to feed on, so they have to make do with what pheasant and hare they can find, supplemented by hen's eggs stolen from cottage gardens! But the one thing you cannot find on the island is folk-singing, usually so prevalent in Dalmatia. The Mljet men will tell you it is because they have no musical ear and can make only a horrible noise!

Coming back in the stillness of the evening to the hotel on the lake, time seems to slip back. In the failing light the building still looks like a living monastery, and the cloisters seem peopled with the ghosts of Benedictine Brothers. To one side of the lake a huge moon makes a path of watery light, while, on the other, a red sunset etches the wooded hill-tops into black silhouette. There is a peace and magic about Mljet which is too precious to be lost.

Šipan, Lopud, Koločep and Lokrum

The group of seven islands near Dubrovnik are known as the Elaphites — a name given to them by Pliny — and the first and biggest is Šipan. Šipan's western end overlaps Mljet, but is much closer to the mainland. The passenger boat calls at Šipanska Luka. It is a fascinating approach. This boat, unlike the large steamers en route for the big islands and Rijeka, keeps close to the rocky spurs of the islets of Olipa and Jakljan. The strata have been laid bare in places by wind and salt — cream-coloured and exciting layers with, at one point off Jakljan, a natural rock-arch close to the water like a beautifully built stone doorway. In contrast, other small spits of land are covered in sharp green maquis. Šipan's port is round a sheltered corner. The island (16½ square kilometres) has two limestone ridges enclosing a Dolomitic valley which produces olives, figs, grapes, almonds, pomegranates and citrus fruit. It is the submerged end of this valley which forms the harbour of Šipanska Luka; there is a second port, Suđurad, formed in the same way, at the other end of the same valley.

The island achieved notoriety during the Roman era when two admirals, Vatinius, belonging to Caesar, and Octavianus, of Pompey's fleet, were fighting a naval battle. Vatinius was in hot pursuit, when Octavianus slipped through the narrow seaway of Prolaz Harpoti, close to the natural rock-arch, and got away. There are the remains of a Roman country villa near Šipanska Luka. During the rich Dubrovnikan period it was fashionable for families to build summer residences for themselves on Šipan; there are some splendid examples including the summer palace of the archbishops. At Renatovo there are the ruins of a small castle where, it is said, René of Anjou (Renato), king of Naples, lived for a time in the fifteenth century. To back this legend up, a coat of arms with the inscription 'Renatus Rex Justus' has been found near by. Šipan is a beautiful little island, quiet enough for a peaceful stay (one small hotel, a *gostionica* and a few private rooms) yet close enough to Dubrovnik to enable one to make leisurely and extensive explorations of that perfect walled city.

The next island is Lopud, separated from Šipan by the Lopud Straits. This island (4·63 square kilometres) has a population of about 400 in the only village, also called Lopud. Like Šipan, it has two limestone ridges enclosing a fertile valley, with submerged ends forming the cove of Sunj in the south-east and Lopud on the wide, north-westerly bay. Although so much smaller than Šipan, Lopud has become a popular tourist resort. It also has the advantage of some fresh-water springs.

There is a broad walk like a promenade close to the sea. On its eastern extremity and slightly higher than the port are the ruins of a Franciscan monastery and cloisters founded in 1483, which was left derelict in 1808. In the sixteenth century it was fortified by a wall and a defence-tower. But the monastery church of Sv. Marija is still in use and has some fine paintings and some very beautifully carved fifteenth-century choir stalls. Although the ancient Greeks settled on Lopud, calling it Delaphodia, and the Romans, who called it Lafota, its really important period came after Dubrovnik had taken it under its medieval republican wing. The population was swelled not only by the well-to-do of Dubrovnik, but by various refugees from the Turks in the fifteenth century. The islanders became famous as seafaring men and took their ships into battle during the campaigns of the Spaniards, fighting off Tunisia, Algeria, Portugal, and even the British Isles. Of the two forts protecting the port, one is called the 'Španjolska Tvrdava', or Spanish Stronghold. Records show that there were between 70 and 80 sea-captains as well as other sailors, and that when the ships were at sea only the women of the town were left. So many men were lost while serving as mercenaries that the expression 'Trista vica udovica' — three hundred widows — was coined with reference to Lopud.

The Dubrovnikan writer Razzi stated that there were two monasteries, thirty churches and a quantity of grand summer residences. Not many have survived, but one ecclesiastical building to remain is the church of the Holy Trinity at the entrance to the harbour. This was built by one of the Dubrovnik sea-captains, Vice Bune (1559 - 1612), and contains his tomb. There are also remains of a Dominican monastery dated 1482. Secular architecture includes the ruined house of a ship-owner, Miho Pracat (1522 – 1607) who must have been active at the time of the Spanish Armada against England in 1588, the restored home of the Đorđić family, and the ruins of the Count's Palace. There is no road, but a small path across the island, called Viktor Dyk walk, leads to the second bay, on the south-eastern side where a fifteenth-century church, Our Lady of Sunj, stands on the site of a twelfth-century chapel and contains some notable paintings including one by Jacopo Palma the Elder and part of a polyptych by Matej Junčić from 1452. Even more remarkable is a carved wooden altar which, it is said, came from a chapel in Westminster after Henry VIII embraced the Anglican faith. A sailor from Lopud bought the altar and gave it to the island.

The remarkable prosperity of this little island came to an abrupt halt in 1667, the year of the great earthquake which, on 6th April, destroyed a large part of Dubrovnik, the town on which it depended.

Nearest to Dubrovnik from the west, and even smaller, is Koločep, again, on the familiar twin limestone-ridge pattern. There are two villages and a total population of 240. This also was known to the Greeks who called it Calaphodia. With the coming of the Slavs the island became part of the province of Travunja, a tribal settlement close to Dubrovnik. The Croat king Ladislas made the island over to the Dubrovnik republic at the end of the eleventh century. In the statute book of 1272, five hundred families lived in the island. Besides agriculture Koločep had a thriving coral industry between the fourteenth and the eighteenth centuries. The fishermen collected the coral not only from the Adriatic but from the Aegean and round the shores of Malta, and sold their products through the merchants of Dubrovnik to the Levant, Italy and France. In 1571, the year of the Battle of Lepanto, the island was fortified at all important points against the attacks of pirates, particularly the dreaded Uluz Ali, sometimes called the Bey of Algiers. One legend says Ali was an excommunicated monk from Calabria who had a particular hate for all church buildings and a compulsive urge to burn them; a pyromaniac of some stature!

To the east of Dubrovnik, and so close to the walled city that it seems within arm's reach, is Lokrum. Densely wooded, it lies across an azure strip of water, facing the old harbour of the city. Small motor-boats make the return trip every ten minutes or so. Once there, visitors are free to wander in the woods, find a small deserted bathing cove, visit the museum, explore the monastery ruins and gardens, order a really first-class lunch in the open-air restaurant, or just fade like chameleons into the general and very beautiful background. The only fixed point is the time of the last boat back to Dubrovnik, as no one is allowed to stay overnight on the island. It is used as much by the people of Dubrovnik as by foreign tourists. It has the full protection of national park status and even smoking is not permitted, though its total prevention would be hard to accomplish.

Benedictine monks built an abbey and church on Lokrum in 1023; the Romanesque basilica with a nave and two aisles was twice enlarged and the bell-tower, which was also useful as a defence tower, remained inside the church area. In the fourteenth and fifteenth centuries a new monastery was built but, sadly, all the buildings suffered in the earthquake of 1667. According to legend King Richard Coeur de Lion landed on Lokrum when he was returning from Palestine after the Third Crusade, probably in the year 1192, after his stay in Corfu. His ship, wrecked in a storm, was washed ashore on Lokrum, the south-western coast of which is steep and barely accessible, though the north-eastern part is lower. The king is supposed to have sworn a vow to build a church in thanksgiving on the spot on which he was saved, though there is no certainty that this was carried out. He did, however, contribute generously towards the building of Dubrovnik's cathedral.

In 1466, at the instigation of Dubrovnik, the monastery on Lokrum became associated with St Justines in Padua, a tactical error on the part of the republic. It gave the Venetians an excuse to claim Lokrum in an effort to quell Dubrovnik's power. In 1631, Venice sent 200 men to the island who immediately set up fortifications. The Dubrovnik republic replied across the narrow strip of water with cannon fire from the fortress of St John on the outer arm of the harbour, and the Venetians beat a hasty retreat. The monks stayed on Lokrum until 1798 when the island was put up for sale and bought by twelve citizens of Dubrovnik. During the Napoleonic wars the French arrived and built a fort on the highest point, calling it Fort Royal. It took the form of a five-pointed star. After the defeat of Napoleon and the subsequent arrival of the Austrians in the area, the fortress was strengthened by the building of 'Maximilian tower'. Finally, in 1859, the island was purchased by the Austrian archduke Maximilian of Hapsburg who built a castle where the destroyed portion of the monastery had been and lived on Lokrum until his departure for Mexico. He also laid out the beautiful park, planting exotic sub-tropical trees and shrubs round the monastery. Maximilian had a tragic life, ending in his execution in Mexico in 1864, but it is said that his happiest days were spent on his little paradise island, sometimes referred to as the 'island of love'. In 1878, the island became the property of Crown Prince Rudolph of Austria, the Empress Elisabeth's son; his life, too, was tragic, and he committed suicide with Mary Vetsera at Meyerling on 30th January 1889. Today the palace has a less fateful atmosphere and houses a fascinating mixture of museum exhibits. Outside there is the enormous rib of a whale, just to put you in the right mood. Inside, the long galleries contain stuffed fishes, birds, and reptiles; a huge terrapin caught near Budva in 1894; stuffed sharks, octopus, sea urchin and squid; and flying fishes of different varieties including the small dark blue one, often seen in the area, called *lastovica*, or little swallow. In another section are examples of sea and land birds, including several species of snipe.

Besides this zoological museum, the Yugoslav Academy of Science and Art has now established the Biological Institute on Lokrum, and there is the Ruđer Bošković Memorial Museum. Outside, in the grounds surrounding the monastery ruins, there is a good collection of new shrubs and trees, recently planted and well labelled in Latin and Serbo-Croat. Varieties of eucalyptus, amaryllis and acacia grow happily side by side with oleanders and other local plants — an extension of Maximilian's garden. But there was a time, between the world wars, when the museum buildings were homes for war orphans, followed later by use as sanatoria for young people. One of the terrible prices paid for long periods of deprivation and malnutrition was a high percentage of tuberculous children. Since the last war a comprehensive health service has made rapid strides towards eradicating diseases of this nature.

Dubrovnik

Across the narrow strait is the fascinating walled city of Dubrovnik, still perfect in its medieval plan, unchanging and indeed unchangeable. The ramparts are unbelievably tall, thick and solid enough for internal passages and stairs, with a parapeted walk on top wide enough for a cart. The walls are still complete, so that it is possible, with two or three hours to spare, to make the circuit of the city at parapet height without once coming down to ground-level. From this vantage point one can witness the whole gamut of twentieth-century life: mini skirts on a washing-line; babies on grandmothers' knees in the sun; a man shaving after his day's work; and, bridging the gap of centuries, an old monk teaching his novices in the shade of some cloisters, or a nun scrubbing at the ancient stone flags of her convent kitchen. Although Dubrovnik suffered severely in the 1667 earthquake, all rebuilding was carried out in the original style in the best Renaissance tradition. The palaces of rich Dubrovnikan families, the monasteries and the churches still stand side by side with the private houses, large and small, of this eternal and still bustling metropolis. The modern hotels with their shining plate-glass are outside the old city, and wheeled traffic is kept out too, so that visitors and inhabitants can walk at will along broad and marble-paved main streets and squares or into the narrow side streets to the boundary walls. There is only one hotel within the walls; it looks on to the market and was once a huge private house of the Plenić family. It has a very small chapel in a terraced garden at first-floor level. One of the Plenić sons, a sea-captain, built the chapel, dedicated to Sv. Antun, for his father who was very old and could no longer go to church. The terrace is now the hotel restaurant, but the door of the chapel stands open and there is still a cloth on the altar.

The people of Dubrovnik are different from any others in Yugoslavia. Subconsciously, their rich patrician past seems to remain with them. The Dubrovnik city-state in the Middle Ages kept free of the rival clutches of Venice and the Ottoman Empire, frequently playing off one would-be tyrant against the other, and waxed rich and powerful in the process.

No amount of detailed reading can take the place of the impact of actually being inside the city and sharing its vivid life. In the soft warmth of the evening, sitting on a terrace or at a wide upper window, the chatter of the city rises to greet one. The swifts and martins, outpacing their slow pigeon brothers, swoop at roof-top height over the small medieval homes etched, layer on layer, before one's eyes. Black silhouettes, these birds dive with a kind of Dubrovnikan daredevil ardour — like élite Black Arrows of some feathered airforce — barely missing the invitingly open windows. The roofs of the buildings are tiled in the special rose-pink and yellow peculiar to Dalmatia, and above them the sky looks like an extra warm Canaletto painting as the sun goes down behind the ramparts. The line of roofs is broken every now and again by cupolas surmounted by delicate filigree iron crosses, like some full and shapely woman's breasts, multiplied and adorned in religion. Bells, some cracked and some musical, chime out, breaking across the chattering of people and of birds. Not only the tourists chatter, but the whole of Dubrovnik, for this is the time of day when everyone emerges. From house, office, or shop, they sit and sip wine in courtyards and gardens, or wander up and down the main streets where paving-stones gleam with the patina of centuries; greeting friends, comparing dresses, eyeing girls, and discussing the latest news.

Dubrovnik has always been a tolerant city, since the days when Epidauron, the Greek (later Roman) colony at Cavtat, was finally sacked by the Avars. The few survivors reached the small island rock, then separated by a seawater creek from the mainland, and settled there to a new life, protecting themselves by some form of early defence walls. This was the beginning of Ragusium, first mentioned by a geographer from Ravenna in the second half of the seventh century and later by the Emperor Constantine Porphyrogenitus in the tenth century. The name Dubrovnik came from a nearby Slav settlement on the mainland and is derived from *dubrava*, meaning a wood or grove. In its present form the name appeared first in a charter issued by the Bosnian Ban Kulin

(1180 – 1204) who, with extraordinary prosperity, developed the trade of his country with the help of the Ragusan merchants. Since those early medieval days the city-state developed, to a remarkably high degree, a conscience for social hygiene and social welfare. Rich families, with working men and girls in their households and estates, were responsible for their well-being, treating them like members of a larger family and providing them with dowries on marriage. To keep the crowded city healthy, anyone coming in through the Pile Gate from the mainland had to wash his hands and feet in the fountain just inside the drawbridge entrance — the famous Onofrio fountain built, with the great aqueduct, in 1438. At the port outside the walls are the lazarettos, built at the end of the sixteenth century. These huge buildings were designed to protect the inhabitants from infectious diseases. All goods arriving in Dubrovnik were liable to disinfection. Today the ancient city port no longer has big shipping traffic, though it still bustles with small craft, private yachts and fishing boats. The big ships go to Dubrovnik's large international port of Gruž, two or three kilometres to the north-west of Pile Gate.

Dubrovnik is a sophisticated city. Besides having on its outskirts some of the best hotels in Europe it has long boasted a night-club, now increased to two; with soft music and even softer lights, one of them nestles inside and underneath the rampart walls. There is an annual six-week summer festival in the city which attracts world-famous orchestras and artists. Concerts and operas are given in the open against fabulous medieval palaces and walls, and Shakespeare's plays are performed in ancient fortresses — unforgettable performances in an unforgettable city.

East of Dubrovnik the chain of Adriatic islands peters out and the few rocky protuberances which break the surface of the sea are insignificant, in size at least, though there are some fascinating points of history attached to some of them. For instance, twenty kilometres from Dubrovnik and offshore from Cavtat there is a group of islets, Supetar, Mrkan, and Bobara, all early Dubrovnikan leper colonies. A Benedictine monastery was founded on Mrkan in 1284, and the ruins of the church are still there. The area was inhabited by Illyrians and later by Greeks (from 689 B.C.), followed by the Romans. There must be many hidden traces of those early times; after all Epidauron, Epidaurus, or Epidaurum — call it what you will — figured as an outstanding centre on the ancient trade route of the old world. It was a point of culture, of riches, and a thriving commercial filter on the long line of communication between Asia Minor, the Baltic and the West.

The only other island proper before the Adriatic coastline takes a sharp turn south past Albania is Sv. Nikola. This is a small pine-covered island overshadowed by the resorts of Budva and Sv. Stefan, both of them islands attached to the mainland by small causeways. Sv. Nikola, close to Budva, has some church ruins, one house, and one small kafana, open at night. The island also produces an endemic lily, *Lilium Buthuense*. The name derives from the Greek for Budva, which was Buthoe.

Budva itself is one of those enchanting, ruggedly walled small cities cemented on to a steep rock rising out of the sea. It has had a rugged history too, and its crooked, narrow streets, with their overhanging houses and shadowy ground-floor store-rooms and cellars, must have witnessed many alarums and excursions in the days of piracy and foreign attack.

Looking at Budva from outside, it is impossible to guess just how many medieval buildings have been crowded in. The old cathedral of Ivan Krstitelj — St. John the Baptist — was built originally during the eighth and ninth centuries, though it has had several reconstructions since. The so-called 'new' cathedral was built in 1418! Budva suffered in the 1667 earthquake and has had other tremors since, lying as it does on one of the earth's volcanic 'paths'. The beaches are made of beautifully coloured stones, in blues, greens, pinks and beige. It is on this part of the coast that the hard white of the limestone rock at last gives place to richer colours.

Facing Budva on the other side of the wide sweeping bay is Sv. Stefan which is even smaller. Sv. Stefan has become famous as a tourist creation since the Second World War, but through the ages it has served as a well-protected homestead for a number of familes. It is said that the first inhabitants were a *Zadruga* called Paštrović who repaired to the islet when danger threatened, storing grain and oil and other supplies. There are still some stone olive-presses. The first building to appear was the chapel of Sv. Stefan in the fifteenth century, followed by another chapel in 1693. Both these are still in good repair at the top of the steep and rocky little hill which rises sheer out of the sea. Both chapels are open, and the older one has some beautiful icons painted on the wood of the altar screen. Often the Serbian Orthodox priest sits outside in the sun, reading a newspaper printed in the Cyrillic alphabet; sometimes he is absent, climbing the mainland mountain to reach his other church, half way up its slope. From the west, as far as Dubrovnik, the Adriatic coast and islands have always been Croatian Roman Catholic, though Dubrovnik has one fine Orthodox church. East of Dubrovnik one slips straight into the Montenegrin/Serbian area of Orthodox religion without at first noticing the change. The same goes for the people. Montenegro, bound up with the Serbian kingdom through the centuries, was as brave as a fighting cock, taking on all comers, whatever their size. But, try as she might, she could not get a share of the Adriatic coast until this century. In the meantime, through the centuries, Dalmatian families settled along the shore, to mingle with Bosnians; so that the feeling 'this is Montenegro' comes only softly to the coastal visitor, whereas inland the impact is immediate. On closer contact, however, there is little doubt. The fierce pride, under the politeness and lavish hospitality of Balkan people, is smouldering there. The willingness to cut off the nose to spite the face is not far away. The certainty that they would rather do things in their own way (and probably better at that) lurks round every corner. One remarkable, and purely Montenegrin, success is the experiment at Sv. Stefan. In 1952, the last remaining ten or eleven fishing families of the rocky promontory were moved to the mainland and rehoused in new modern villas. Then two architects, Radanović and Maximović, began converting the entire rockbound village into one luxury hotel, though the original idea was to form an artist's retreat. The cottages were turned into apartments, and the steep twisted lanes and steps leading from one height to the next were left intact for guests to stroll in. The restaurant and bar terraces were built on a high rock platform overlooking the sheer drop to the sea below, so that guests can feed the fish which swarm in the clear waters without even getting up from their tables. There is natural fresh water on Sv. Stefan syphoning under the sea, from the mainland mountain; it only has to be piped to each apartment bathroom. A perfect setting, one would say, in the middle of some of the finest scenery in the world.

Perhaps the best approach is by ship from the sea. A small boat carries passengers from the liner to the landing-stage. Hotel maids look down from small defensive windows to view the newcomers — just as pirates' wives might have gazed on invaders. The approach up the granite cliff is the old rough-hewn path of the seamen. Though Korčula, Budva and Dubrovnik may have their moulded ramparts and towers, Sv. Stefan would have no truck with elegance; the ramparts are living rock with rough walling in between, built only where necessary to complete the protection. Inside the cottage walls the apartments have been well designed; only the bathrooms — the last word in 1960 when the hotel was opened — strike one as a little old-fashioned, though blending with the cottage atmosphere. Yugoslavia now has brilliant tiles and mosaics in her hotels and cruising ships, and perhaps Sv. Stefan could afford to upgrade hers. Moving up and down the steep streets from an apartment to the restaurant affords the visitor a pleasant pre-prandial excursion, good for the liver, though one does hear some elderly Americans, who live their lives by elevators, complaining at times that they cannot walk another flight of steps before dinner. Most of the inmates of Sv. Stefan disappear during the day, either to sunbathe on their private apartment patios or to the mainland for excursions or bathing. The narrow causeway, which formed itself gradually from the sand and shingle washed there by contrary waves and winds, offers two splendid sides from which to laze and swim, and all traffic is barred at the mainland end. On the mainland, the secluded and lovely 'Queen's beach' almost faces Sv. Stefan; Queen's Beach is attached to the Miločer Hotel which has extensive and beautifully tended gardens and was formerly the royal summer palace. From the mainland Sv. Stefan looks its best — a fairy castle rising, on its rock, from the dark blue sea. The stone of the buildings is as creamy white as elsewhere in the Adriatic, but the mortar used between is red like the land near by and gives added colour and warmth to the structure.

In many ways it is perfect, perhaps too perfect. And because there are no village families left on the rock — not even an ordinary village shop — its perfection becomes a little like an opulent film-set. The hotel is one of the most expensive in Yugoslavia; it is beautifully run, very efficient, equipped with materials of the finest quality, but without the eagerness which makes other centres an experience of contact and joy. The top employees seem to worry unduly about presenting the bills, while the bottom ones worry about their own lack of money. In short, the spirit of the centuries seems to have disappeared, and for me it is a paradise *manqué*. If only they had left one or two indigenous families there, one or two tiny shops, instead of installing each peasant-fisherman in a brand new mainland villa, complete with bathroom! Then indeed Sv. Stefan might have been the paradise that was so carefully planned — and so carefully lost — instead of the cardboard fantasy which is proving a home from home for film stars and starlets; and the original idea of preserving an artist's retreat might even have been included.

From Sv. Stefan it is only about 34 nautical miles to Ulcinj and the Albanian frontier; and the next island in the Adriatic is Greek Corfu.

Chapter 12 **Corfu**

The island of Corfu, relatively far from the other Adriatic islands, and the only one belonging to present-day Greece, lies at the narrow sea gates between the Adriatic and the Aegean and is the most northerly of the seven Ionian islands. It has a splendid climate, and the colours of the flowers and the trees — olives, oranges, and lemons — measured against the fantastic blues of sea and sky, give the island a quality almost of unreality. It is a place of paradox: at once hard and soft, lush and poor, sophisticated and simple. Although so bound up historically with the other Adriatic islands, its outlook is very different. In some respects it has its feet on the ground — aided perhaps by fifty years of British governorship — but in others it has a 'never-never land' quality. Geographically it is much softer and more lush than the Yugoslav islands. Only at its northern tip, where the craggy mountains overhang the deep dark sea and the hills are terraced, does it resemble the mainland coasts of Dalmatia and Montenegro. Certainly, in looks and temperament, it bears no resemblance to Albania whose grey, bare mountains overshadow the east coast of Corfu, at their nearest point less than a mile distant.

Shaped like a sickle, it is long and slim, its width varying between 4 and 30 kilometres. But its length of 70 kilometres allows great variation between north and south. The south is soft and comparatively low with gentle slopes and sandy shores, while in the north the heights and cliffs are steeper, the scrub at the top of the mountains sparser, and the colours more violent.

The interior of the island is like a continuous Elysian country scene. A network of lanes, mostly with indifferent, or worse, surfaces, leads from one tiny colour-washed village to the next. The warm pinks and pale greens of cottage walls are an indication of Corfu's independence of spirit, not conforming with the brilliant white which is uniform to most of Greece. Dogs, slim cats, and sloe-eyed children look up at the few passing cars; small boys wave a greeting. The little island donkeys, tethered to shady posts, shiver their flanks and make superior faces, knowing that four sure feet are better than rubber wheels when it comes to cobbled steps in village streets. This part of Corfu cannot have changed much in a thousand years. The cottages are tremendously strong with great square pillars holding up patio arches, and some of the older olive trees were striplings at the time of the Crusades. Some of the hamlets are perched on steep hillsides overlooking the broad central valley. This valley, where most of the grain and vegetables grow, is extremely fertile, and water has never been a problem. Yet there is a half-plan to annex the area for tourism, replacing essential food with golf- and race-courses. One hopes that the Corfiot term 'tomorrow', which really means 'the year after next', will prove an insurance policy strong enough to save their agriculture — any island's necessity.

In the interior it is easy to believe the ancient legends. One of the earliest names for Corfu was Drepanum — the Greek word for a scythe — a reference to its shape. Legend says that the Greek goddess Demeter hid her scythe each evening under Corfu before going up to Mount Olympus for the night. Odysseus too, according to legend, met Nausicaa, the beautiful and determined daughter of King Alcinous, on Corfu after being shipwrecked on the shore and staggering naked into the arms of Nausicaa, to the horrification and terror of the princess's attendants. Nausicaa spoke sharply to them and managed to introduce him to her father, the king, who did not know at the time that Odysseus was a married man! Eventually, however, the hero was sent packing — to Nausicaa's disgust. Later he lost his ship which is supposed to have been turned to stone by the tetchy Poseidon who, with Zeus, 'had it in' for poor Odysseus. Today the small ship lies offshore from Corfu in the guise of a small island. Jason and Medea — the pair who may have landed at Osor on the island of Cres — also called at

Corfu, according to Appollonius of Rhodos. It has also been suggested that Corfu — 'Rich scarf to my proud earth' — is Prospero's island in Shakespeare's play *The Tempest*, though without much foundation. But the island needs no legends to boost its charm; there is enough of paradise in its make-up to satisfy any twentieth-century traveller, though strangely enough archaeological exhibits are sparse. Undoubtedly, below the surface, there is a wealth of material waiting to be discovered; as yet the surface has hardly been scratched.

From legend to history is a short step in these areas. Corcyra, known sometimes as White Korkyra to distinguish it from Black Korkyra (Korčula), was founded in 734 B.C. when the Corinthians formed a colony there. But, 'biting the hand that founded them', the Corcyreans fought a naval battle with the Corinthians in 665 B.C., beating them soundly. In 434 B.C. there was another naval battle, off the south-east tip of Corfu, when the Corcyreans and the Athenians joined against the Corinthians. This battle marked the beginning of the famous Peloponnesian War.

As so often happens rival factions started quarrelling among themselves, weakening the strength of Corcyra against outside enemies. Between 361 and 229 B.C. there were civil wars on the island until the Romans, seeing their chance to step in, took over Corfu in 229 B.C. and stayed in command until A.D. 733: 962 years in all. No wonder Cato and Cicero and others, as well as Tiberius and Nero, were attracted to this delightful green and pleasant land!

This period was not, however, without troubles. Various raiding parties arrived, attracted by its riches. Among recorded raids are one by 'Vandals' in A.D. 445 and trouble with the Goths under Tottila in 562.

In 733 Corfu first came under the religious guidance of Constantinople instead of the Roman Pope. From the first millennium until 1386 (when Venice finally took over at Corfu's request) the island was a pawn in the big-time politics and struggles for power of the period. In 1080, and again in 1203, it was temporarily annexed by the Sicilians. In 1205 the Venetians arrived, only to be driven out by their constant enemies, the Genoese. A year later, the Venetians retook Corfu, but in 1214, eight years later, the island united with Epirus. It was at this time that the castle of Angelocastro, on the heights above Paleokastritsa, was built. In 1259 Michael II of Epirus metaphorically packaged up the long-suffering Corfiots and gave the island to King Manfred Hohenstaufen of Sicily as a dowry for his daughter Helen. Eight years later Charles d'Anjou, king of Sicily and Naples in that most complicated period of Western history, took possession. The island remained under this dynasty until the death of Charles II in 1386, though it was frequently raided not only by Venetians but also by Catalans and Aragonese.

In fact history has rubbed familiar shoulders with Corfu over a period of at least 3,000 years. Even in the last 200 years its changing masters have left a fascinating debris of architecture, especially in the town. In 1797 Corfu was taken over by the French under General Gentilly, after Napoleon had destroyed the Venetian republic. The Venetians had been in Corfu since 1386, invited there by the Corfiots to restore order out of chaos after the confusion of the previous century. They did a surprisingly good job in many ways, putting up many beautiful buildings, and it was during this time — when the Greek mainland was dominated by the Turkish Empire — that Corfu became rich as the centre of trade between East and West.

Under the French a new street plan was laid out and the celebrated arcaded buildings on the Esplanade were created. These resemble those of the Rue de Rivoli and were, perhaps, a nostalgic reminder, like the horse-drawn landaus still in Corfu today, of the delights of Paris for the French garrison.

This first period of occupation only lasted two years — the French were driven out by a Russo-Turkish force in 1799 — but during these two years (when Bonaparte was an army general without royal aspirations) the famous Livre d'Or, instituted by the Venetians, was destroyed. Corfiots had never bothered about such distinctions as 'nobility' or 'heraldry' but, when the Venetians colonized, their method was to leave the government in the hands of a few 'noble' local families, raising them to a kind of peerage and recording them in the Golden Book. Napoleon's republican zeal did not, however, cause the final destruction of the book, though its pages were burnt at the time. When the French returned in 1807 at the ratification of the Treaty of Tilsit, Napoleon had given himself royal status, and so the Livre d'Or was brought out into the open again with his blessing.

When the British arrived in 1814 after Napoleon's defeat at the Battle of Leipzig, they settled in for exactly half a century. Corfu became the capital of the new 'Ionian republic' which included the six other Ionian islands under British suzerainty. The British, as always, were thorough and put into operation the town water-supply, which is still in use, as well as some very creditable public works and buildings.

The more imposing part of the town where the British mark is still most noticeable, unlike the enchanting, higgledy-piggledy muddle of the old, Venetian section, centres round the Esplanade. This is a huge open space slightly reminiscent of an old-style cantonment exercising square, though with various objects relieving the barren surface; a British 'folly', classically colonnaded, and a fountain or two together with some statues. On the seaward side and standing at the same cliff-top level as the Esplanade and old town is the old fortress or citadel, divided from the town by a sea-water moat and spanned by a bridge. Fortified by huge vertical walls which rise from sea-level, it still bears the mark of Venice in the form of the familiar winged lion. Opposite the fort, on the landward side of the square, are the 'rivoli' arcades, known as 'The Liston'; under them centres Corfu coffee-life; and it is a small idiosyncratic fact that the 'old' families frequent only a particular section of the long line of café tables!

On the north side of the Esplanade is what is now called the Royal Palace. It was built between 1819 and 1823, to serve as the official residence of the British Lords High Commissioner. A splendid Georgian building of lovely proportions, it has a colonnaded walk along its length, some noble archways and, above the two tiers of Georgian windows, a decoration of seven medallions to represent the seven Ionian isles: Bellerophon on the winged horse, Pegasus, for Lefkas; Cephalus the Hunter for Cephalonia; Zakinthos, with lance in hand, for Zante; the head of Odysseus for Ithaca; a trident for Paxos; Aphrodite and a dolphin for Cythera; and a rudderless ship for Corfu, itself. One wonders how significant this last sign is; it is easy to relate it to Corfu's indepence of spirit, her quiet way of paddling her own canoe, and her tolerance of the individual's caprice and even eccentricity.

During the British occupation the Residence (built of limestone imported from Malta) was also the seat of the Order of St. Michael and St. George which was created primarily to reward service in Malta and the Ionian islands. The Offices of Government were also housed in the building which became the official palace when Corfu and the other islands were ceded to Greece on the accession of King George I of Greece. It remained the official palace until the Junta's coup in 1967, though it fell into a state of disrepair after the privation and bombing of Corfu during the last war. In 1956 it was restored and now houses an interesting Sino-Japanese collection of art. Other reminders of the British time in Corfu are the Ionian Parliament House, the Peristyle of General Sir Thomas Maitland (who was the first British Lord High Commissioner) and a bronze statue of Lord Guilford who founded the Ionian Academy. There is also the splendid villa with the unlikely title of 'Mon Repos'. This large house, built at the same time as the Residence, is south of the Esplanade on the road to Kanoni. It was the birthplace of Prince Philip, Duke of Edinburgh, and was the summer holiday home of the Greek royal family until 1967. 'Mon Repos' is not open to the public, but, for those with enough energy to climb down the steep steps to sea-level near the 'Mon Repos' bathing beach at Kardaki, there is an entrancing decision to be taken — to taste, or not, the water which trickles out of the rock face. The water is clear and diamond-bright in the hot sun, but there is an inscription in Greek by Lorenzo Mabili let into the cliff face. It reads:

> 'And so must some god have ordained,
> That any stranger who there wets his lip,
> Must never to his fatherland return.'

To risk it or not? The island of Corfu houses many foreigners, none of them anxious to leave! British residents alone number more than 50, with over 100 part-timers. Some of them are famous, some are not; but all come for the gentleness of climate, for the good manners, and the beauty, of the island.

Social organizations in the town are numerous. There is the famous Corfu Reading Club, established in 1836. It contains a fascinating collection of miscellaneous books, frequently added to by bequests, and is housed in a lovely Venetian building with an external stone staircase and arcaded balconies on its upper storeys. Here not only reading takes place but social gatherings and bridge drives. Other societies include the Old Philharmonic Company, and the New, the Touring Club, Alpine Club, two cricket clubs and tennis and football clubs.

Corfiots are particularly musical, and there are, in all, at least six wind (or concert) bands. The oldest of these is the St. Spyridon Philharmonic Society, founded in 1840 in a typically Corfiot way. One of the many feast days falls each year on 11th August when the casket containing St. Spyridon is carried in procession round the town in memory of the miracle by which, it is said, the Turks were driven away. The Turkish fleet had drawn near when, during a great storm, they saw a black figure holding a fiery brand in the sky. Naturally, the Turks fled. The townspeople wanted some music to accompany the procession, and they applied to the British High Commissioner, Lord Douglas, to borrow the military band. He, however, refused to lend it for what he considered an unsuitable purpose. So five or six young Corfiots scratched up a wind band of their own, complete with a uniform in the blue and red Ionian colours and brass helmets.

British rule in Corfu seems to have left no bitterness. It is true that it is now more than a hundred years since it ended, and it is also true that the authoritarian rule of the first High Commissioner was not much relished. But on the other hand, 'stone' ginger beer and apple chutney have passed into Corfiot cookery books, while cricket — the introduction of which is doubtfully attributed to Lord Byron — is still played, complete with 'bollerr', 'overrs' and 'scorrs'! The Corfiots do not forget, either, that Edward Lear, who was a serious painter as well as a light-hearted writer, lived in Corfu for six years. His attractive, almost mystical, pictures are reproduced with great affection. Many of the islanders speak excellent English — a language they would rather remember than Italian. Corfu, like some of the other islands, suffered greatly under Italian occupation.

Perhaps the most Greek things in Corfu are the Orthodox churches and monasteries, though even these have a special Corfiot 'softness' about them which matches their priests, who seem more gentle than their austere black habits and tall hats might suggest. They bask in the sun, sitting in wicker chairs, their wives and grandchildren clustered round them. It always surprises me to remember that Greek priests marry, though unlike their lay friends they only marry once, it seems. In the island, other men can have several chances without incurring local displeasure, though anything over three marriages is considered excessive!

The oldest church still standing is dedicated to Saints Jason and Sosipatros. It is a pure Byzantine building dedicated to the saints who first brought Christianity to the island in the second century A.D.[1] Records show that this church was restored in the eleventh century, and it contains, on a wall, the oldest icon in Corfu, also eleventh century.

Father Demetrius lives quietly beside the church of Jason and Sosipatros with his family, his cats and their kittens, and his brilliant flower garden. He owns a well from which he will proudly offer a glass of 'the purest water in Corfu'! Over a cup of thick, strong Greek coffee — the word 'Turkish' is not used in Greece — he talks hopefully of the union of the churches.

Another Corfiot priest finds it hard to resist a pretty face — in the nicest possible way! One Holy Week the Director of Tourism was asked if a Dutch girl model could pose in front of some icons. The request was passed on to the priest. 'Curses on you, my son,' he replied, 'for putting such temptation in my way during Holy Week.' But the pretty face won, and the model was photographed in the church.

The church of St. John, near the cathedral in the old town, houses a rare collection of icons. In fact, it was the original cathedral, and in existence in 1520 — as far back as records go — before the present cathedral was built. The oldest icon in St. John's, dating from 1520, is near the screen, always a solid structure in Orthodox churches. The picture is an example of 'linear' art: a series of small pictures, bordering the main one, denoting the life

[1] Some local sources make it as early as A.D. 65.

of St. John. Nikiforus Theotokis, famous in Russia, came from St. John's church. Brilliantly clever as a theologian, mathematician, astronomer and physicist, he went to Moscow and became a bishop and President of the Academy. He died in Russia in 1800. The church of St. John was very small and, to accommodate all the townspeople, a sort of annexe was included on one side, and the pulpit was placed on the right of the altar instead of the left so that all the congregation could see. It owns a very beautiful Bible, dated 1712, whose covers are of intricately worked gold, a particularly Ionian art form. Every Orthodox church, when it is consecrated, must have a relic — perhaps a sliver of holy bone or a splinter of the Cross. The priest, Father John, will sometimes show the jewelled container which holds the fragment of wood — St. John's consecration relic.

The cathedral, Panaghia Spileotissa (the Virgin of the Cave) was built in 1577. It contains some icons and neo-Byzantine paintings and, in the crypt, the headless body of St. Theodora, wife of the Byzantine Emperor who reigned from A.D. 829 to 842.

In the middle of the crowded 'old town' is the church dedicated to St. Spyridon. It always seems full of people. Corfiots appear just to 'pop in' on their daily rounds. Most of them visit the small side chapel containing the mummified body of St. Spyridon in its ornate silver casket. Little bits of gold and silver, old jewellery and candles are everywhere — small gifts of thanksgiving for favours asked for and received. There is an atmosphere of affection rather than hushed reverence here, and even the most sophisticated schoolchildren pause to kiss the saint's silver container while asking his help with exams. Four times a year the saint leaves his chapel to be carried in procession, in the inner gold casket, on thanksgiving days: Palm Sunday for deliverance from plague in 1629; the Saturday before Easter for the relief from famine; 11th August for deliverance from the Turks in 1716; and the first Sunday in November after another bout of plague in 1673.

Corfu town and its surroundings hold enough interest for weeks of sightseeing. Among other things, the Old Citadel on the twin-peaked rocks to the seaward side of the Esplanade should not be missed. Called Corypho in the Middle Ages, it gave its name to Corfu and was first fortified by the Byzantines after a Slav raiding party had been repulsed in 933. The present rugged fortifications were built by the Venetians, while the clock tower and barracks — still in use — are a legacy of British rule. The 'Old' citadel, by the way, is not to be confused with the 'New', whose construction was begun, also by the Venetians, in 1572. The massive walls surrounding the old town were built after the Turks attacked Corfu in 1537. Inside these thick defences the townspeople could rely on greater safety for the houses they built. Some of these dwellings are seven or eight storeys high and constructed as flats, often for the use of different sections of the same family, and are still used as such today. It is this old section of the town which now offers a splendid tourist shopping area. A very beautiful Venetian well, the Kremasti Well, is tucked away in a quiet square of the church of 'The Hanging Virgin' — the name thought to be a reference to a picture or statue. The well, given in 1699, was the first public water supply. There is a legend which says that, should disaster threaten Corfu, a young calf appears from the well with chains on its legs.

A fascinating morning can be spent in the newly arranged archaeological museum, to the south of the town towards the bay of Garitsa. It gives a good idea of just how far back into antiquity Corfu goes. The museum is very modern, and this contrast with the exhibits heightens the impact. The most important archaeological find on the island, the west pediment of the temple of Artemis or the Gorgon, is in a large end room. This Doric temple dates from 600 – 580 B.C. From the proportions of the pediment it is estimated that the building was 48 metres by 23, and 17 metres high. It was built at the time when Corfu (Korkyra) was breaking its associations with Corinth. In the middle of the pediment is Artemis, the 'beautiful goddess of the forests'; her belt consists of two serpents rising at the front to form an emblem like the sign of Aesculapius. Her two sons, Pegasus and Kryssaorus, are near her with, on either side, a panther. The concentric circles representing the panthers' fur — beautiful, simple, and graphic — are from the old Babylonian art which influenced the Corinthians. Titan and Zeus are also represented. Titan is throwing a thunderbolt while Zeus tries to stop him, and they are holding each other's hair in a wrestling pose. Among the decorations are the key pattern and the ancient 'fishbone' relief. A good deal of the original is missing, but restoration has been cleverly done so that all the original is in its bold relief and detail, while the missing parts are indicated in flat diagram.

Besides this almost overpowering exhibit there is part of a 'zoophorus', or animal frieze, part of a statue, perhaps of Achilles and a fragment which may be Athena. There are coping tiles, some large and Corinthian, some Lakonian or Spartan. It is interesting that original tiles were made of pottery while replacements were of stone. The museum holds some fascinating IOUs made in lead — recent finds from the period 600 – 550 B.C. — in the form of flat rectangles, about 4 inches by 2; some curious flat round wine jars, prehistoric remains of 3000 – 2000 B.C.; and examples of the early 'furrow writing', which runs up and then down the next line like the track of bulls or oxen pulling a plough. Among the most recently found, but earliest, exhibits are some massive funeral urns; four or five feet high, they were discovered near the town.

In spite of its great age, the island does not strike one as an archaeological trust. On the contrary, it is very much a twentieth-century mecca for holidaymakers. Most of the hotels are on the east coast, north and south of Corfu town, though there are a few on the west side and more are planned. The town of Corfu itself has eighteen hotels, with one or two very 'de luxe' among them, and plans for six more. In fact the whole town is a delicious mixture of extreme sophistication and old-world charm.

One of the delights of the island is that although there are many tourist centres, many good hotels, and an island population of 110,000, it seems in some magical way never to be overcrowded. Outside the town the small tavernas strung along the coast road keep their village charm despite the holiday invasion. The huge groves of olive trees on the slopes above the sea remain sylvan and uncluttered by summer bric-à-brac. The great sweeps of fine sandy beach still manage to give the impression that Pan, with his horns and his cloven feet, is just around the corner — eccentric, slightly comic but, above all, benign.

A drive by coach or car along the road, built by the British 11th Regiment R.E. in 1828, to Paleokastritsa is a 'must'. The route winds through huge olive groves to the west coast, and it is worth by-passing the direct route to the twin bays and driving up the steep mountain slope to Lakones, directly above Paleokastritsa, first. From the jutting terrace of the small café the view is fantastic. Three hundred and fifty metres below lie the narrow curved bays with, on a small promontory between, the tiny ochre-coloured monastery. The rich yellow of the monastery walls contrasts almost savagely with the aquamarine and sapphire of the sea and its edging of white foam on the rocks. Out to sea are two small islands, one of which looks like Odysseus' ship turned to stone. Paleokastritsa is one of the sites for the legend of Nausicaa and her father King Alcinous. Even higher than Lakones, on a pinnacle above, are the remains of the thirteenth-century Byzantine fortress, Angelocastro. The energetic can climb to this, though it takes some time. Most people make the descent from Lakones to the bays of Paleokastritsa, paying the monastery a visit, and then enjoying lunch by the sea. There are hotels and terraced restaurants by the sea which, in this area, has a treasure-island quality. It is also famous for lobsters though, unfortunately, shell-fish are no less expensive here than elsewhere in the Mediterranean!

There are many other lovely excursions, for the entire island is bounded by beautiful and secluded, often people-less, bays. There is Ai (or Ayios) Gordis, best approached by a drive through the hills, and villages such as Kastellani. The road is terrible, but the countryside and the flowers make it well worth while. Ai Gordis has an almost deserted bay of fine sand and peacock-coloured water with a small restaurant run by Theodor Dukatis. The cooking is good, and he has a few simple rooms to let with outside washing facilities, rather like that of an autocamp. Glyfada, more sophisticated and also on the west coast, is near.

On the east side, travelling north of Corfu town, are first Gouvia, then Dassia (complete with Club Mediterranée), Ipsos, Pyrghi, Barbati and Nissaki, all with hotels either existing or projected. And if you want a change from shining modernity the Castella, turned into an hotel from an Italian nobleman's 'folly', is as full of antiques and curios as a baronial film set. The vast mock-medieval house was built on a hill, ten minutes away from the sea, in the 1880s on the few remains of a fifteenth-century fortress. The park and trees are magnificent; so is the view and the cooking; and the battlements, complete with hot-oil-pouring slots, look most realistic.

Kassiopi, closest to the coast of Albania, is in the far north. It is the site of a temple of Zeus and has the ruins of a Byzantine fort. Nero is said to have played the harp and sung here, while Tiberius is reputed to have stayed in the vicinity; but before visiting Kassiopi or Sidari, also in the North, it may be necessary to get a police permit. Not far away is Mount Pantocrator, the highest mountain (914 metres), with one of the oldest monasteries at the top. The Brothers are very hospitable, and the view over the island and to Albania is superb. To be there at the time of the religious festival (3rd to 6th August) is a rare experience.

South of Corfu town, strung along the sandy coastline towards the tail of the island, are Benitses and Messonghi. The several hotels in this area have their feet almost dabbling in the sea. Hotel rooms open on to beautifully kept flower gardens only a few feet from the water. Messonghi is flat and mysterious with a kind of 'bamboo' atmosphere; at the end of the long stretch of sand a slow river moves into the sea, cutting off the village beyond from the hotel area. Between Messonghi and Corfu the road runs along the coast, and each village has its local tavernas. The one I remember best is Charlie's Bar. The simple tables and chairs allow space for Corfu dancing — a particularly sedate brand of Greek dancing — to juke-box music which plays bouzoukia or jazz indiscriminately. Greek dance is liable to start at any time, though it is only the Greek men who dance, never Greek women. This does not apply to tourists, however, who are all roped in, girls, matrons and self-conscious men. But the *pièce de résistance* at Charlie's comes when his son chooses a table loaded with glasses and bottles, picks it up in his teeth, and glides round the floor to the music without dropping or breaking a glass. At this taverna I saw another example of Corfu's tolerance towards human frailty. An old man with a gentle face and misty blue eyes picked a carnation and came across the dance area to present it — silently. He melted away again, and I was told he was the village thief — a title conferred on him in the same way as our 'village idiot' might be. He had once broken into a house to steal, but the owners came back too soon and he rushed into an attic to hide. Finally, after hours of silent waiting, nature overtook him and he was forced to urinate on the floor. It trickled through the old boards and he was discovered and sent away to be reprimanded: a thief *manqué*, tolerated by the rest of the village.

One of the most attractive of the old olive estates lies in this part of Corfu, with its lands gently sloping down to the sea. The house, built about two hundred years ago, is mellow and rambling with huge reception-rooms and a visitors' book which reads like a cross between a royal diary and *Who's Who*. One of the attractions for royal visitors was to take tea on the balcony which has a perfect view of the blue sea.

In the days when families were large and gala nights frequent and glittering, entertainment in this old house must have been very lavish. In a cupboard occupying the whole of one wall in the butler's pantry, there is a 600-piece dinner-service presented to the family by the British Governor after he had stayed with them for two months while his Residency was being built in 1819.

Across a small farmyard where pedigree cows and their calves are tethered in the shade is the door of the small and beautiful Orthodox church, belonging to the house. St. Stefano is full of icons, some very beautiful. There are Russian ones presented by visitors in the days of the Czars, covered in worked silver except for the faces of the Virgin and Child, with silver points for haloes. There are many old and valuable books and a solid screen with a curtained doorway before the altar. Women are never allowed beyond the screen, as the Orthodox Church believes that the devil lives in woman. The perpetually burning lamp suspended before the screen in the church is of ornate silver. Water, as well as oil, is put in, as the water, sinking to the bottom, prevents the silver from becoming too hot and melting. The flame is lit at the top of a flower-cup or sepal which floats on the oil; these are collected in the woods and dried, and have the appearance of wood sorrel.

In Corfu, where there are between four and five million olive trees and where olive oil forms 50 per cent of the island production, the fruit is left on the tree until it falls naturally instead of being picked. The trees are huge and some are known to be nearly a thousand years old. As soon as the fruit has fallen it must be gathered quickly before it has time to become acid. Up to five degrees of acidity is acceptable, but any more makes the oil bitter and strong. When the fruit is left to ripen, the trees only bear every second year. On the other hand, if olives are picked every year the tree must be pruned hard to keep it manageable and small, while picked olives, being less ripe, produce less oil, so that Corfu does not lose by her 'softer' way of cultivation. Olive-presses can be found in all parts of the island; some are ancient and made of stone, unchanged since Corinthian times.

The oldest of the 'big' houses are found in the interior. Of such an age is the home of the Curcumelli family, 'Afra'. This fifteenth-century house, still lovely in its faded state, seems almost in suspended animation, like some sleeping Aurora. Originally it was a monastery, and there are still cloisters at ground-level. Among the more distinguished of the family members was Sir Demetrius Curcumelli, K.C.M.G., who was Regent of Corfu in 1864 when the British left. It was he who presented the Farewell Address — a panegyric of goodwill — when the garrison departed. Now the house is almost empty at times, but to take tea there is like re-entering the Edwardian era. Sponge sandwiches gleam on the dark polished wood of the dining-table while conversation floats round in perfect English. The walls show spots of damp from the old roof, but the silver and china cabinets are beautifully kept, and the gun-room is a treasure trove of past and present weapons and curios from the East.

How different a building is the celebrated Achillion Palace, built for the Empress Elisabeth of Austria. She had been to Corfu as a young and beautiful, if dissatisfied, woman in 1861. Twenty years later she asked for a 'summer retreat' — a kind of Petit Trianon — to be built for her: 'a Phaeacian Palace with pillared colonnades and hanging gardens, safe from prying eyes; a palace worthy of Achilles'. The local consul was entrusted with the work but, overawed by the grandeur of the task, he produced a huge baroque confection, with decorated walls and painted ceilings, heavy marble columns and gilded finery everywhere, while statues — undistinguished and brought from an Italian villa — were placed overwhelmingly in the gardens. The house is on three floors reached by a fantastic staircase. The view from the balconies at the top is superb and, as it is all built into the side of the steep hill, the top floor also has a lovely garden with huge palm trees and shrubs, roses and lily pools. Now that it is all turned into a casino it seems, at last, to have found its true *métier*. There are gaming tables on all floors with an outstanding restaurant, both inside and in the garden, at the top. The food is excellent, the atmosphere international, the dancing not too crowded and the management carried out by cosmopolitan as well as efficient Germans who will sit and chat at one's table in French, German, Greek, or English, with such facility that they don't seem to know that they have changed languages!

On the ground floor of the Achillion is a small 'house' museum, devoted to things collected by Elisabeth of Austria and, later, by Kaiser Wilhelm who bought it ten years after Elisabeth was assassinated in Geneva. Among Elisabeth's belongings are fans, photos and curios; a silver cigar case with a diamond motif and a crown was a present, perhaps, from 'Uncle Nicky' of Russia. There is some furniture and a heavy silver tray with the dolphin and crown. The dolphin was her Corfu emblem (in honour, perhaps of Achilles' mother, Thetis), and the empress ordered dolphins' heads with everything. All the door-handles, bath-taps, and even terraces are flanked by dolphins. Perhaps the most poignant relic is one of a collection of coins in a glass case. It is the third on the right and is gold. This coin is one of those given by the Emperor Maximilian of Mexico — the man who build his palace on Lokrum outside Dubrovnik — to each of the soldiers who formed his firing squad in 1867. Giving his coins, he asked the soldiers to fire at his heart and not his face; he was a good-looking man. One of the firing squad kept his coin and returned it to the Empress Elisabeth, who was Maximilian's sister-in-law.

When the German Kaiser bought the villa it took on a very Germanic air. His bedroom was heavily furnished and there was a saddle in his study as a desk-chair. His court painter (before the time of official photographers) has left various rather bad paintings depicting Wilhelm gazing across at Albania from the balcony, or Wilhelm reviewing Corfu from his yacht. Each year he arrived in April with the entire court, plus presents for the 'natives'. One year he brought gifts of soap which did not increase his popularity. But perhaps what offended the cultivated taste of the Corfiots most was the monster statue of Achilles which he erected in the garden. Enormous, prefabricated, and complete with an electric light bulb at the top, the inscription read 'From the greatest German to the greatest Greek'. The French hastily tore it down in the First World War while the place was a military hospital! Poor frustrated and physically handicapped Wilhelm! He did leave one spot in Corfu which is still enjoyed today; Kaiser's Seat, at the top of steep hills overlooking the sea near the village of Pelekas provides the perfect position for watching the sun setting across the Adriatic.

The flowers in Corfu are fantastic, whether cultivated round cottage doorways, in the grander gardens, or wild. In June wild white roses cover the hedges at the side of the road, and there are mauve monkshood as big as cultivated ones. Near Paleokastritsa dark pink and white cystus, with flowers two inches across, riot on the mountains. There are dianthus and wild sweet-smelling thyme, iris and all the beautiful shrubs of the Mediterranean.

It is a perfect setting for King Alcinous's 'fair city'; but where, indeed — if at all — was it? Although many historians agree that Corfu was probably Scheria, the land of the Phaeacians, the only part of the island to reveal any very early remains — except for a few Minoan potsherds near Paleokastritsa — is just south of Corfu town, mostly in the grounds of the 'Mon Repos' palace. It was here that the ruins of the temple of Artemis were found. Menecrates' cenotaph, too, is in this area (dating from the seventh century B.C.) as well as the later Roman baths and an early Christian basilica, obviously built from the stone of much earlier buildings, including one block decorated with the 'fish-bone' motif. The nearby sheltered inlet of Kanoni, with the small Pondikonissi island, fits Homer's story as easily as Paleokastritsa, except for the prevailing sea currents which flow (though slowly) the wrong way if he was sailing from Malta, or Melita. But would any ancient Greek sail across the open Mediterranean? It was not in their scheme of things; they sailed within sight of land. And in Homer's story Odysseus, King of Ithaca, would be more likely to visit the islands off the Adriatic coast which were well known to the ancient Greeks. Perhaps it is possible that Odysseus, like St.

Paul in later years went, not to Malta but to that other Melita now known as Mljet. Then his route home would have hugged the coasts of what are now Yugoslavia and Albania in his southerly course towards Ithaca. The north-east Adriatic wind in the storm would then have driven his ship to the eastern shore of Corfu.

So although at a cursory glance Corfu might not seem to be closely linked with the other islands in the Adriatic, ever since pre-history the connections have been there. Perhaps the essential difference between Corfiot and Yugoslav islander is that the great Slav migration of the seventh and eighth centuries failed to spread as far as Corfu, so that the directness of the Slav approach has never been fused on to the Greek stock of the Corfiot, as it was in the islands which Corcyra colonized about five centuries B.C. But history has linked all the islands with many incidents. Richard Coeur de Lion, who landed on Lokrum near Dubrovnik, stayed in Corfu in 1192. The Corfiots, like Serbs and Croats, suffered greatly at the hands of the Turkish Empire, one terrible incident being the siege of August 1537 when the Turks landed in Gouvia Bay. Before they were driven off 15,000 islanders were taken and sold into slavery — nearly half their population. Yet another link was the Battle of Lepanto. It was fought just south of Corfu at Patras, and 1,500 Corfiot sailors took part. Remembering the long years of Ottoman oppression, the people of Corfu must have been overjoyed when, after a heroic fighting retreat across the mountains of Albania, the remnants of the Serbian army were received by the British in Corfu, finally ending 500 years of Turkish domination. In 1918 the kingdom of Yugoslavia was proclaimed in Corfu's Municipal Theatre.

What of the character of the people? First, they are tolerant though perhaps guarded in their initial encounters, but once contact has been made, they are gentle, kind, warm-hearted, transparently eager and fantastically generous. Secondly, they take great pride in their beautiful island, conscious that it has long been the holiday playground of rich Europeans. They are conscious, too, that the Corfiot Greek is not quite like other Greeks — how could he be? But though independent of spirit, they are not like the Yugoslav islanders; not so clear and sharply etched. They are as ebullient as any Greek yet not so extrovert. The boys in the schools, between classes, are as vociferous as any ancient Greeks may have been and, perhaps like them, have an instinctive knowledge of their own nature allowing their 'year after next' attitude to save them from rash decision, and preserve the unique qualities of their island.

Corfu, like the other Adriatic islands, has a thriving tourist industry. The modern international Corfu Port, with its new expansion, has ample room for all vessels from small sailing yachts to the biggest cruising liners, and receives visits from Yugoslav passenger ships. Do the people of Corfu ever remember that their ancient Greek forebears peopled the Yugoslav islands to the west? Perhaps not; but the shared characteristic of making strangers welcome still exists, oiling the wheels of understanding on both sides and overcoming any barriers of language or frontier.

Works consulted

Naši Otoci Janranu. Dr. Ivo Rubić. Slobodna Dalmacija. Split. 1952.

Kulturna Historija Hrvatske. Zvane Črnja. (English translation by Vladimir Ivir.) Office of Information, Zagreb. 1962.

A Short History of Yugoslavia. Ed. Stephen Clissold. Cambridge University Press. 1968.

Yugoslavia. A. W. Palmer. Oxford University Press. 1964.

Jugoslavija Turistička Enciklopedija. Ed. Ljubica D. Trajković. Beograd. 1958.

The Greeks Overseas. John Boardman. Penguin Books Ltd. 1964.

Corfu. (Venus of the Isles. Ed. John Forte.) 'A Synoptic History.' Dr. Theodore Stephanides. East Essex Gazette Ltd. 3rd edition. 1968.

Pavijest Otoka Brača. Dasen Vrsalović. Brač. 1968.

Kulturni Spomenici Otoka Brača. Andre Jutronić. Brač. 1968.

Srednjovjekovni Statut Bračke Komun. Anton Cvitanić. Brač. 1960.

Lastovski Spomenici. Cvito Fisković. Split. 1966.

NiN. Mate Suić, Šime Batović, Janko Belošević. Zadar. 1968.

Pag. Mate Suić. Zadar. 1963.

Ancient Pula. Stefan Mlakar. Arheološki Musej, Istre, Pula. 1965.

Amfiteatar u Puli. Stefan Mlakar. Arheološki Musej, Istre, Pula. 1966.

Vis. I and II. Dr. Grga Novak. Zagreb. 1961 – 5.

Grapčeva Špilja na Hvaru. Dr. Grga Novak. Zagreb.

Viški Spomenici B. Gabričeuić. C. Fisković, N. Božanić-Bezić and A. Cvitanić. Split. 1968.

Thesis on Hvar. Dr. Niko Duboković-Nadalini.

Useful information

JAT (Jugoslovenski Aerotransport) fly from the following international airports: Amsterdam, Athens, Beirut, Berlin, Bratislava, Brussels, Budapest, Cairo, Copenhagen, Frankfurt, Graz, Istanbul, London, Milan, Moscow, Paris, Prague, Rome, Sofia, Stockholm, Tripoli, Tunis, Venice, Vienna, Warsaw, Zurich.

Civil airports within Yugoslavia include: Belgrade, Dubrovnik, Ivangrad, Ljubljana, Mostar, Ohrid (S), Priština, Pula, Krk, Sarajevo (S), Skopje, Split, Titograd, Tivat, Zadar (S), Zagreb, Zabljak (S).

(S) = seasonal.

JADROLINIJA the Yugoslav shipping line which runs the coastal cargo/passenger ships of the Yugoslav Adriatic, most of the car-ferry services and a number of luxury cruising ships in the Mediterranean and elsewhere, is based on Rijeka. The ships in the Adriatic run daily, on a regular time-table basis, and it is possible to work one's way from island to island and port to port stopping off here and there and catching the next ship round in the time schedule. This can also be done on a cruising basis from country to country in the Middle East. All the bigger ships have dining-saloons and cabins, and the company has a reputation for extremely good food.

YACHTS

Foreign yachts are free to sail in Yugoslav coastal waters provided they are properly registered in their country of origin and possess the necessary travel papers for their passengers and crew. On arrival in Yugoslav waters foreign yachts and motor-boats must register immediately with the port authorities, submitting two copies of their itinerary at one of the following ports of entry:

Koper; Umag; Piran; Pula; Raša; Rovinj. All on the coast of Istria.

Rijeka; Mali Lošinj; Zadar; Senj; Šibenik; Split; Ploče; Korčula; Dubrovnik. All on the Dalmatian coast.

Zelenika; Hercegnovi; Bar. Between Boka Kotorska and the Montenegrin/Albanian frontier.

Customs formalities and health control take place at the port of entry. The Harbour Master's office is always situated on, or very near, the quay. In case of damage or bad weather yachts can put up in any port, and many places on the islands and the coast are equipped to carry out both major and minor repairs. There are refuelling facilities at all major points and, in a number of ports, special fuel pumps for yachts including: Mali Lošinj, Crvena Luka (near Biograd), Split, Korčula, Vela Luka, Hvar, Dubrovnik.

A number of ports are able to provide winter berths and moorings for visitors wishing to lay up their craft in the Adriatic for the winter.

For CORFU entry regulations are the same and, on registering, the yacht-owners will be handed a Transit Log, valid for one year. The large harbour attached to Corfu town has all yachting facilities.

N.B. In Yugoslavia all jetties, whether attached to private houses or not, can be used for tying up in case of necessity.

TOURIST AGENCIES

In Yugoslavia there are a number of tourist agencies which have offices in all tourist resorts. These are separate organizations which work in competition with each other. They organize excursions, book accommodation, supervise private room accommodation and act as information offices.

RECOMMENDED READING

Yugoslavia. Phyllis Auty. Thames & Hudson. 1965.

Black Lamb and Grey Falcon. Rebecca West. Macmillan 1942. Reprinted 1967.

Eastern Approaches. Fitzroy Maclean. Jonathan Cape. 1949.

The Companion Guide to Yugoslavia. J. A. Cuddon. Collins. 1968.

FOR YACHTSMEN

The Adriatic. A sea-guide to its coasts and islands. H. M. Denham. John Murray. 1967.

Index of place names, and historical and other figures

N.B. Where names constantly recur throughout the book, for example Illyria, Greece, Rome, Venetians, Dalmatia, Croatia, etc., they have not been included in this index.